Organic Rankine Cycle for Energy Recovery System

Organic Rankine Cycle for Energy Recovery System

Special Issue Editor

Andrea De Pascale

MDPI • Basel • Beijing • Wuhan • Barcelona • Belgrade • Manchester • Tokyo • Cluj • Tianjin

Special Issue Editor
Andrea De Pascale
Department of Industrial Engineering,
University of Bologna
Italy

Editorial Office
MDPI
St. Alban-Anlage 66
4052 Basel, Switzerland

This is a reprint of articles from the Special Issue published online in the open access journal *Energies* (ISSN 1996-1073) (available at: https://www.mdpi.com/journal/energies/special_issues/ORC_Energy_Recovery_System).

For citation purposes, cite each article independently as indicated on the article page online and as indicated below:

LastName, A.A.; LastName, B.B.; LastName, C.C. Article Title. *Journal Name* **Year**, *Article Number*, Page Range.

ISBN 978-3-03936-394-0 (Pbk)
ISBN 978-3-03936-395-7 (PDF)

Contents

About the Special Issue Editor

Andrea De Pascale is Associate Professor of Fluid Machinery and Energy Systems at the University of Bologna, Italy. He teaches master courses on Environmental Impact of Energy Systems, Advanced Energy, Systems, and Fluid Power Systems. As a mechanical engineer, he obtained his PhD at the University of Bologna, where he is currently working at the Energy System Group of the Department of Industrial Engineering. His scientific activity deals with thermodynamics of advanced energy systems, advanced gas turbines, CHP and micro-CHP systems, power-to-gas, renewables, and waste heat recovery technologies. He is responsible for a laboratory on technologies for micro-cogeneration. Within this lab, research is currently carried out by implementing test benches and developing numerical and experimental studies on micro-generators, with special reference to the ORC technology. He has authored of more than 100 publications in international journals and congresses, and he is a referee for many scientific journals. He has been awarded the "John P. Davis Award" by the ASME in 2015 for a technical study on gas turbines integrated with ORC in off-shore application.

Preface to "Organic Rankine Cycle for Energy Recovery System"

The rising trend in the global energy demand poses new challenges to humankind. The energy and mechanical engineering sectors are called to develop new and more environmentally friendly solutions to harvest residual energy from primary production processes. The Organic Rankine Cycle (ORC) is an emerging energy system for power production and waste heat recovery. In the near future, this technology can play an increasing role within the energy generation sectors and can help achieve the carbon footprint reduction targets of many industrial processes. In particular, there are still many un-used hot streams available for recovery in various stationary power generators for civil and tertiary applications and in several highly intensive industries. Additional applications can come from the transportation sector, where waste engine heat in heavy vehicles and ships can be used to achieve fuel savings. Moreover, low-enthalpy flows from renewable sources can be exploited in thermodynamic cycles based on the Rankine architecture. The ORC is already a well-proven option in large-size plants, but not all technological aspects are currently solved/optimized; the state of the art still requires cost-effective improvements in order to enlarge the market opportunities. Meanwhile, the ORC is still developing in small-scale and/or micro-generation applications, where efficient and low-cost ORC components are not ready for the market yet and problems must be solved. This Special Issue focuses on selected research and application cases of ORC-based waste heat recovery solutions. Topics included in this publication cover the following aspects: performance modeling and optimization of ORC systems based on pure and zeotropic mixture working fluids (Andreasen et al.); applications of waste heat recovery via ORC to gas turbines and reciprocating engines (Carcasci et al., Branchini et al., Valencia et al.); optimal sizing and operation of the ORC under combined heat and power and district heating application (Branchini et al.); the potential of ORC on board ships and related issues (Baldasso et al.); life cycle analysis for biomass application (Stoppato et al.); ORC integration with supercritical CO2 cycle (Espinel Blanco et al.); and components for small ORC, including proper design and related internal fluid issues (Casari et al., Fadiga et al.). The current state of the art is considered and some cutting-edge ORC technology research activities are examined in this book.

Andrea De Pascale
Special Issue Editor

Article

Assessment of Methods for Performance Comparison of Pure and Zeotropic Working Fluids for Organic Rankine Cycle Power Systems

Jesper Graa Andreasen, Martin Ryhl Kærn and Fredrik Haglind *

Department of Mechanical Engineering, Technical University of Denmark, Nils Koppels Allé Building 403, DK-2800 Kgs. Lyngby, Denmark; jgan@mek.dtu.dk (J.G.A.); pmak@mek.dtu.dk (M.R.K.)
* Correspondence: frh@mek.dtu.dk

Received: 29 March 2019; Accepted: 6 May 2019; Published: 10 May 2019

Abstract: In this paper, we present an assessment of methods for estimating and comparing the thermodynamic performance of working fluids for organic Rankine cycle power systems. The analysis focused on how the estimated net power outputs of zeotropic mixtures compared to pure fluids are affected by the method used for specifying the performance of the heat exchangers. Four different methods were included in the assessment, which assumed that the organic Rankine cycle systems were characterized by the same values of: (1) the minimum pinch point temperature difference of the heat exchangers; (2) the mean temperature difference of the heat exchangers; (3) the heat exchanger thermal capacity ($\bar{U}A$); or (4) the heat exchanger surface area for all the considered working fluids. The second and third methods took into account the temperature difference throughout the heat transfer process, and provided the insight that the advantages of mixtures are more pronounced when large heat exchangers are economically feasible to use. The first method was incapable of this, and deemed to result in optimistic estimations of the benefits of using zeotropic mixtures, while the second and third method were deemed to result in conservative estimations. The fourth method provided the additional benefit of accounting for the degradation of heat transfer performance of zeotropic mixtures. In a net power output based performance ranking of 30 working fluids, the first method estimates that the increase in the net power output of zeotropic mixtures compared to their best pure fluid components is up to 13.6%. On the other hand, the third method estimates that the increase in net power output is only up to 2.56% for zeotropic mixtures compared to their best pure fluid components.

Keywords: organic Rankine cycle system; zeotropic mixture; heat exchanger; low grade heat; thermodynamic optimization; method comparison

---

1. Introduction

The organic Rankine cycle (ORC) power plant is a viable technology for conversion of heat to electricity. The heat-to-electricity conversion is enabled by circulation of an organic working fluid in a closed thermodynamic cycle. When the temperature of the heat input is low or the electrical power output of the plant is low, the ORC system features advantages compared to the steam Rankine cycle, since the working fluid properties of organic fluids are favorable over the properties of steam in these applications [1–4].

A way to improve the system efficiency when utilizing low-temperature heat is to use a zeotropic mixture as the working fluid [5–7]. Zeotropic mixtures change the temperature during the phase change, which is opposed to the isothermal phase change process of pure fluids. The temperature difference between the saturated vapor and liquid temperatures is typically denoted as the temperature glide. By employing zeotropic mixtures as working fluids in ORC units, it is possible to utilize the temperature glide to reduce the temperature difference during heat transfer in the primary (heat input) heat exchanger and condenser. On the other hand, the use of zeotropic mixtures is often related to larger heat exchangers due to degradation of heat transfer performance [8], and lower mean temperatures of the condenser and primary heat exchanger.

The performance of pure fluids and zeotropic mixtures have previously been compared based on a wide range of performance indicators employing various modeling methods. The methods employed are typically thermodynamic optimization or economic (thermoeconomic or technoeconomic) optimization. In thermodynamic optimization, the objective function can, for example, be the thermal efficiency, exergy efficiency, or the net power output. The modeling detail is typically restricted to flow sheet level including energy and mass balances. This approach has been used extensively for preliminary fluid selection. When used for selection and comparison of pure fluids and mixtures, a value for the minimum pinch point temperature difference in the heat exchangers is typically fixed and assumed equal for all fluids. When evaluated based on this approach, there is a general consensus in the scientific literature that the zeotropic mixtures provide significant thermodynamic benefits compared to pure fluids [6,9–14]. For a 120 °C geothermal heat source, Heberle et al. [9] found that i-butane/i-pentane $(0.8/0.2)_{mole}$ achieves 8% higher second law efficiency compared to pure i-butane. Andreasen et al. [10] identified 30 high performing fluids based on net power output for two heat sources and found that 19 fluids were zeotropic mixtures for a 120 °C heat source while 24 were zeotropic mixtures for a 90 °C heat source. Lecompte et al. [11] reported a 7.1–14.2% increase in the second law efficiency for zeotropic mixtures compared to the corresponding pure fluids for a 150 °C heat source. For heat source temperatures ranging from 150 °C to 300 °C, Braimakis et al. [12] compared the performance of five hydrocarbons and their mixtures, and found that zeotropic mixtures achieved the highest performance for all the investigated temperatures. The performance gains are generally attributed to improved temperature profile matching in the condenser [6,10] resulting in reduced exergy destruction (or irreversibilities) [9,11,13]. Another general conclusion from such studies is that the ORC units using mixtures require heat transfer equipment with larger capacities ($\bar{U}A$ values) [9–12] and heat transfer areas A [13] when compared to pure fluids. Thereby, it remains unclear whether the increased performance compensates for the larger investment required for the heat exchangers when zeotropic mixtures are used as working fluids.

Baik et al. [15] fixed the total cycle UA value and optimized a transcritical ORC unit using R125 and three subcritical ORC units using R134a, R152a and R245fa. The results suggest a 5% larger net power output of the transcritical cycle compared to the highest performing subcritical cycle (R134a). In a comparison of pure and mixed working fluids for ORC units, Baik et al. [16] fixed the total cycle heat transfer area assuming tube-in-tube configurations for heat exchangers. Based on this comparison, they concluded that the use of zeotropic mixtures did not have a significant impact on the performance of the condensation process for transcritical ORC units. Bombarda et al. [17] compared the performance of an ORC unit and a Kalina cycle unit for diesel engine waste heat recovery based on equal logarithmic mean temperature differences in the heat exchangers, and found that the two cycles obtained similar performance. The works by Baik et al. [15,16] and Bombarda et al. [17] focus on specific case studies, and do not include an assessment of the employed methods or the implications of selecting those methods.

The trade-off between increased investment in heat transfer equipment versus improved thermodynamic performance can be accounted for in economic optimizations of ORC units. The models employed in economic analyses are typically based on a flow sheet level model for determining the thermodynamic states in the process, which is combined with economic models for estimating economic performance criteria, for example the net present value or levelized cost of electricity. In the economic models, the cost of the ORC unit is typically determined from equipment cost correlations usually involving sizing of heat transfer equipment. Previous studies employing thermoeconomic or technoeconomic optimization methods for fluid comparisons do not clearly indicate whether mixtures or pure fluids are more feasible to use [18–25]. Le et al. [18] minimized the levelized cost of electricity for n-pentane/R245fa mixtures, and found that pure n-pentane yielded the lowest values. A similar conclusion was reached by Feng et al. [19] in a comparison based on a multi-objective optimization of levelized cost of electricity and exergy efficiency. In an other study, Feng et al. [20] found that a mixture of R245fa and R227ea was unable to reach lower levelized cost of electricity than R245fa. For a case study based on waste heat recovery, Heberle and Brüggemann [21] minimized the system cost per unit exergy for R245fa, i-butane, i-pentane, and i-butane/i-pentane, and found the lowest values for pure i-butane. Oyewunmi and Markides [22] investigated the thermoeconomic trade-offs for binary mixtures of n-butane, n-pentane, n-hexane, R245fa, R227ea, R134a, R236fa, and R245ca, and also found that pure fluids were the most cost-effective. On the other hand, Heberle and Brüggemann [23] demonstrated a 4% reduction in the electricity generation cost for propane/i-butane compared to i-butane for utilization of a geothermal heat source at 160 °C. Based on multi-objective optimizations of exergy efficiency and specific ORC unit investment cost, Imran et al. [24] found improved economic and thermodynamic performance for R245fa/i-butane $(0.4/0.6)_{mass}$ compared to pure R245fa and i-butane. Andreasen et al. [25] found that the outcome of the performance comparison between R32, and R32/R134a (0.65/0.35) depends on the amount of investment made. At high investment costs, the mixture R32/R134a (0.65/0.35) obtained higher performance than R32, while the two fluids reached similar performance at low investment costs. A major drawback of using the economic optimization methodology is the high model complexity and computational time required compared to thermodynamic methods. For this reason, thermodynamic methods are preferred for fluid screenings where many fluid candidates are considered.

As indicated above, contradicting conclusions are resulting from thermodynamic and economic optimization regarding the feasibility of using zeotropic mixtures. This is especially the case when the minimum pinch point temperature differences are assumed to be equal for all fluids. Thus, there is a risk that fluid screenings based on thermodynamic optimization and fixed minimum pinch point temperature difference assumptions might identify economically infeasible fluids. This indicates a need for improved methods for fluid screening, enabling effective identification of economically feasible zeotropic mixtures.

Alternative thermodynamic methods have been proposed and discussed in relation to vapor compression cycles [26,27]. McLinden and Radermacher [26] proposed a method for specifying the total heat exchanger area per unit capacity, and claimed that this method provides a fair basis for comparing the performance of pure fluids and mixtures. Högberg et al. [27] assessed three methods for comparing the performance of pure fluids and mixtures in heat pump applications. The first method was based on equal minimum approach temperatures (equivalent to equal minimum pinch point temperature differences), the second was based on equal mean temperatures in the heat exchangers and the third on equal heat exchanger areas. They concluded that the first method should be avoided, while the third method is the preferred method. The second method was evaluated good enough for rough performance estimations.

In the framework of preliminary performance evaluation of zeotropic mixtures and pure fluids for ORC systems, an assessment of available methods for modeling the heat exchangers is relevant, since the same values of the minimum pinch point temperature differences result in higher performance and larger heat transfer equipment for zeotropic mixtures [9–13]. Therefore, it should be considered whether the use of zeotropic mixtures results in higher performance when the same size of heat transfer equipment is used for the zeotropic mixtures and the pure fluids. Since the assumption of the same minimum pinch point temperature differences does not result in the same size of heat transfer equipment, it is relevant to consider alternative methods for modeling the performance of the heat exchangers in ORC systems. The relevance of an assessment of methods for thermodynamic performance evaluation of working fluids, is supported by the strong preference towards fixing the minimum pinch point temperature difference presented in the scientific literature. Generally, there is a lack of a quantitative assessment of the implications of selecting this approach in comparison to alternative options. Such an assessment is particularly relevant in the case of performance comparison of pure fluids and zeotropic mixtures for ORC systems, since significant thermodynamic performance benefits have been estimated for zeotropic mixtures compared to pure fluids for the same values of minimum pinch point temperature differences [6,9–14], while economic optimizations have resulted in contradicting conclusions regarding the feasibility of using pure fluids and zeotropic mixtures [18–25].

The objective of the present study was to quantify the influence of different modeling methods on the results of the thermodynamic performance comparison of working fluids for ORC systems. The analysis considered methods which are relevant at an early stage in the ORC unit design procedure when many working fluid candidates are considered as possible alternatives. The conventional method of assuming the same minimum pinch point temperature differences in the heat exchangers for pure fluids and zeotropic mixtures was compared to three alternative methods for specifying heat exchanger performance in thermodynamic modeling of ORC systems. The four methods considered in this study are: (1) assuming the same minimum pinch point temperature differences for all working fluids; (2) assuming the same mean temperature differences for all working fluids; (3) assuming the same thermal capacity values ($\bar{U}A$ values, the product of the overall heat transfer coefficient, $\bar{U}$, and the heat transfer area, A) for all working fluids; and (4) assuming the same heat transfer areas for all working fluids. First, it was considered how the net power outputs of ORC units using the working fluids propane, i-butane, and two mixtures of propane/i-butane (mole compositions of 0.2/0.8 and 0.8/0.2) vary as a function of the condenser size, represented by the minimum pinch point temperature difference, the mean temperature difference, the thermal capacity ($\bar{U}A$) value, and the heat transfer area, respectively. Subsequently, a fluid ranking considering 30 working fluids (pure fluids and mixtures in subcritical and transcritical configurations) was made using the following methods: (1) assuming that all fluids have the same minimum pinch point temperature differences in the condenser and primary heat exchanger respectively; and (2) assuming that the sum of the primary heat exchanger and condenser thermal capacity ($\bar{U}A$) is the same for all fluids. Based on the results of the analyses, the methods were assessed and the feasibility of using zeotropic mixtures in ORC power systems was discussed.

The major novelty of the paper is that it provides a quantitative assessment of using different methods for specifying heat exchanger performance in the thermodynamic comparison of zeotropic mixtures and pure fluids for ORC power systems. Such an assessment has not previously been carried out in relation to ORC power systems. In comparison to the method assessments for vapor compression cycles presented by McLinden and Radermacher [26] considering subcritical mixtures of R22/R114 and R22/R11, and Högberg et al. [27] considering subcritical mixtures of R22/R114 and R22/R142b, we extended the analysis by including a fluid ranking considering 30 different pure fluids and zeotropic mixtures comprising both subcritical and transcritical configurations. The conclusions obtained are not only relevant for ORC power systems, but also apply for other thermodynamic processes, as discussed in the paper.

Section 2 describes the implementation of the different models and methods. Section 3 presents and discusses the results of the analyses and includes the method assessment. The paper is ended by Section 4 where the conclusions of the study are presented.

2. Methods

The ORC unit analyzed in this study consisted of an expander, a condenser (cond), a pump and a primary heat exchanger (PrHE) (see Figure 1). The heat input to the unit was provided by a hot water stream, while the heat released from condensation of the working fluid was transferred to a cooling water stream. The case of liquid heat source and sink fluids justified the use of counter-flow heat exchangers, which enabled the utilization of temperature profile matching in the primary heat exchanger and condenser. The mechanical power generated by the expansion of the working fluid was transferred to a generator, enabling the ORC unit to deliver a net power output ($\dot{W}_{net}$) defined by the difference between the power consumption of the pump and the expander (neglecting electrical and mechanical losses):

$$\dot{W}_{net} = \dot{m}_{wf}[h_3 - h_4 - (h_2 - h_1)] \tag{1}$$

The numerical simulation models were developed in Matlab® 2018b [28] using the commercial software REFPROP® version 9 [29] for working fluid property data, and the open source software CoolProp version 4.2 [30] for properties of water. The assessment and comparison of the methods was demonstrated based on a case assuming a hot fluid inlet temperature of 120 °C. A detailed comparison of the net power output variation as a function of the minimum pinch point temperature difference, the mean temperature difference, the $\bar{U}A$ value, and the heat transfer area of the condenser, respectively, was carried out. The working fluids selected for the detailed comparison were, propane, i-butane, and two mixtures of these fluids, which have previously showed promising performance in subcritical ORC systems [10,21]. The mole compositions of the two mixtures were selected to be 0.2/0.8 and 0.8/0.2, since these compositions result in a temperature glide around 5 °C, corresponding to the selected temperature rise of the cooling water. Subsequently, the performance of propane/i-butane was compared with the 29 other fluids (pure fluids and mixtures in subcritical and transcritical configurations) identified by Andreasen et al. [10].

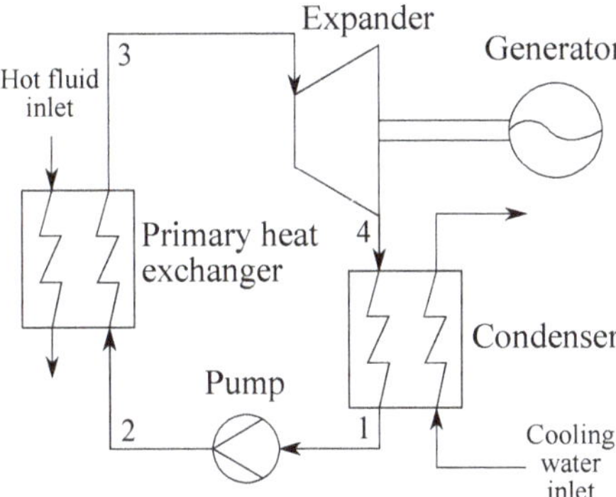

Figure 1. A sketch of the organic Rankine cycle system.

2.1. Thermodynamic ORC Process Model

The thermodynamic models described here provide the basis for the four different methods compared in this study. The only difference between the three methods employing constraints on the minimum pinch point temperature differences (ΔT_{pp}), the mean temperature differences (ΔT_m), or the $\bar{U}A$ values of the heat exchangers, was the calculation of the heat exchanger performance variable (i.e., ΔT_{pp}, ΔT_m, or $\bar{U}A$ was calculated and constrained). The fourth method, which based the performance comparison on heat exchanger surface areas, needed to be supplemented by models for dimensioning of heat exchangers (see Section 2.2).

The modeling conditions used for simulating the ORC unit were based on Andreasen et al. [10] and are shown in Table 1. Additional assumptions were as follows: no pressure loss in piping or heat exchangers, no heat loss from the system, and steady state condition and homogeneous flow. The decision variables were optimized to maximize the net power output of the ORC system, while respecting the constraints on primary heat exchanger and condenser performance, and the minimum expander outlet vapor quality. The optimization was carried out in two steps, where the first optimization step was carried out by running the particle swarm optimizer available in Matlab with a population of 30,000 for 50 iterations. The best solution found in the first step was used as the starting point for Matlab's pattern search optimizer. In Table 1, the primary heat exchanger and condenser constraints are denoted as minimum pinch point temperature difference constraints, however these constraints were substituted by constraints on maximum values of $\bar{U}A$ depending on which method was applied. Note that the heat exchanger parameters were implemented as constraints and not as fixed parameters. However, by optimizing the primary heat exchanger pressure, degree of superheating, working fluid mass flow rate, and the condensation temperature, the optimum values of minimum pinch point temperature differences or $\bar{U}A$ values converged to the limiting values.

Table 1. ORC unit modeling conditions.

Parameters and Variables	Value/Range
Fixed parameters	
Hot fluid temperature (°C)	120
Hot fluid mass flow rate (kg/s)	50
Hot fluid pressure (bar)	4
Cooling water inlet temperature (°C)	15
Cooling water temperature rise (°C)	5
Cooling water pressure (bar)	4
Condenser outlet vapor quality (-)	0
Pump isentropic efficiency (-)	0.8
Expander isentropic efficiency (-)	0.8
Constraints	
Min. pinch point temp. diff., primary heat exchanger [*] [°C]	10
Min. pinch point temp. diff., condenser[*] (°C)	1–10 [**]
Min. expander outlet vapor quality (-)	1
Decision variables	
Primary heat exchanger pressure (bar)	1–$0.8 \cdot P_{crit}$
Degree of superheating (°C)	0–50
Working fluid mass flow rate (kg/s)	5–200
Condensation temperature at bubble point (°C)	17–35

* Results presented in Section 3.2 employ heat exchanger constraints based on the $\bar{U}A$ value rather than minimum pinch point temperature differences. ** The condenser minimum pinch point temperature difference was varied in steps of 1 °C.

For the primary heat exchanger and condenser, a counter-current flow heat exchanger configuration was assumed. The location of the minimum pinch point temperature difference in the primary heat exchanger was assumed to be at the inlet, outlet, or the saturated liquid point. The temperature difference in the condenser was checked at the working fluid outlet (bubble point) and at the dew point. Calculations of $\bar{U}A$ values and mean temperature differences were done by discretizing the primary heat exchanger and the condenser in $n = 10$ control volumes. In the discretization, it was ensured that the bubble and dew points were always located on control volume boundaries. The total $\bar{U}A$ values of the heat exchangers were calculated by summing the contribution from each control volume $(\bar{U}A)_j$:

$$\bar{U}A = \sum_{j=1}^{n}(\bar{U}A)_j = \sum_{j=1}^{n}\frac{\dot{Q}_j}{(\Delta T_{lm})_j} = \sum_{j=1}^{n}\left[\frac{\dot{m}_c(h_{c,o} - h_{c,i})}{(T_{h,o} - T_{c,i}) - (T_{h,i} - T_{c,o})} \cdot \ln\left(\frac{T_{h,o} - T_{c,i}}{T_{h,i} - T_{c,o}}\right)\right]_j \tag{2}$$

where $\dot{Q}$ is the heat transfer rate, ΔT_{lm} is the log mean temperature difference, T is temperature, subscript j refers to control volume j, subscripts c and h refer to the cold and hot side of the heat exchanger, and subscripts i and o refer to inlet and outlet of control volume j. The log mean temperature correction factor was not included in the expression, since the heat exchangers were assumed to enable counter-current flow. Counter-current flow is required in order to utilize the temperature glide of zeotropic mixtures for temperature profile matching with the hot fluid and the cooling water, and can be achieved with plate, tube-in-tube, and single-pass shell-and-tube heat exchangers.

The mean temperature difference was calculated based on the following equation:

$$\Delta T_m = \frac{\dot{Q}}{\bar{U}A} \tag{3}$$

2.2. Shell-and-Tube Heat Exchanger Model

A shell-and-tube heat exchanger model was used for estimating the heat transfer area of the condenser. The model was used for estimation of the heat transfer areas presented in Section 3.1.4. The heat exchanger was assumed to have one tube pass and one shell pass (see Figure 2).

The heat transfer surface area was calculated based on the following equation:

$$A = \pi d_{ou} L N_t \tag{4}$$

where d_{ou} is the outer tube diameter, L is the tube length, and N_t is the number of tubes.

The tube length required for transferring the required heat was calculated by discretization of the heat exchanger:

$$L = \sum_{j=1}^{n} L_j = \sum_{j=1}^{n} \frac{(\bar{U}A)_j}{(\bar{U}A)'_j} \tag{5}$$

where $(\bar{U}A)'_j$ is the $\bar{U}A$ per length of tube, which was calculated as:

$$\frac{1}{(\bar{U}A)'_j} = \frac{1}{\alpha_{in,j}\pi d_{in} N_t} + \frac{\ln(d_{ou}/d_{in})}{2\pi\lambda_t N_t} + \frac{1}{\alpha_{ou,j}\pi d_{ou} N_t} \tag{6}$$

where d_{in} is the inner tube diameter, λ_t is the thermal conductivity of the tube material, $\alpha_{in,j}$ is the inner tube heat transfer coefficient (hot side), and $\alpha_{ou,j}$ is the outer tube heat transfer coefficient (cold side).

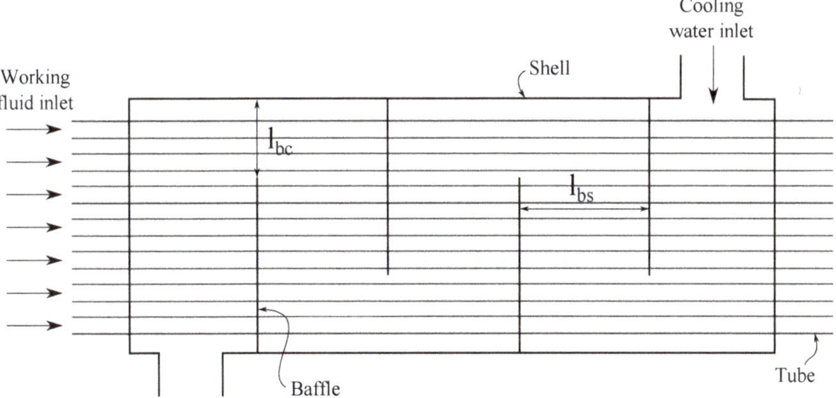

Figure 2. A sketch of the shell-and-tube condenser.

In the heat exchanger model, the number of control volumes was equal to 30. In case the working fluid was superheated vapor at the inlet to the condenser, the condenser was sized to perform both the desuperheating and the condensation of the working fluid. An overview of the implemented heat transfer and pressure drop correlations is provided in Table 2. The modeling conditions assumed for the condenser are listed in Table 3. The tube diameter, tube pitch ratio and baffle cut ratio were selected to represent commonly used values according to the guidelines provided by Shah and Sekulić [31]. A low value of the shell bundle clearance corresponding to a fixed tube sheet design [31] was selected. This enabled a design without sealing strips, since there was no need to restrict the bypass flow between the shell inner wall and the tube bundle. The tube-to-baffle hole diametral clearance and the shell-to-baffle diametral clearance values were selected based on the values used in Example 8.3 in Shah and Sekulić [31]. The thermal conductivity of the tube material was selected to represent stainless steel [31]. The pressure drops were fixed to ensure comparable pumping power for all considered ORC systems. The values of pressure drop were selected to ensure that the flow velocities of the liquid in the shell and the vapor in the tubes were within the limits specified by Coulson and Richardson [32]. The tubes were arranged in a 30° triangular configuration to enhance heat transfer performance [31,33] and no tubes were placed in the window section in order to minimize tube vibration problems [31]. Shell side heat transfer and pressure drop correction factors accounting for larger baffle spacing at the inlet and outlet ducts compared to the central baffle spacing were neglected.

The model of the shell-and-tube heat exchanger was previously presented and verified by Andreasen et al. [25] and Kærn et al. [34]. The implementation of the Bell–Delaware method [35–37] was verified by comparison with the outline presented by Shah and Sekulić [31]. The cross-flow, leakage flow, and by-pass flow areas were predicted within 0.11% (discrepancies were due to rounding errors) of the values reported in Example 8.3 in Shah and Sekulić [31]. The shell side heat transfer coefficient and pressure drop for single phase flow of a lubricating oil were predicted within 0.15% (discrepancies were due to rounding errors) of the values reported in Example 9.4 in Shah and Sekulić [31]. The implementation of the condensation heat transfer correlation was verified to be within 0.8% of the predicted heat transfer coefficients for i-butane presented in Figure 5 in Shah [38]. Discrepancies can be attributed to inaccuracies in obtaining data points from the figure.

Table 2. Shell-and-tube heat exchanger model overview.

Model	Reference
In-tube flow	
Single phase heat transfer correlation	Gnielinski [39]
Single phase friction factor	Petukhov [40]
Condensation heat transfer correlation	Shah [38]
Mixture effects in condensation	Bell and Ghaly [41]
Single phase pressure drop	Blasius [42]
Two-phase pressure drop	Müller-Steinhagen and Heck [43]
Shell-side flow	
Pressure drop and heat transfer correlations	Bell–Delaware method [31,35–37]
Ideal cross flow heat transfer correlation	Martin [44]

Table 3. Condenser modeling conditions.

Parameter Description	Value
Tube outer diameter, d_{ou} [mm]	20
Tube inner diameter [mm]	16
Tube pitch ratio, p_t/d_{ou} [-]	1.25
Baffle cut ratio, l_{bc}/d_{ou} [-]	0.25
Shell bundle clearance [mm]	11
Number of sealing strips [-]	0
Tube-to-baffle hole diametral clearance [mm]	0.8
Shell-to-baffle diametral clearance [mm]	3
Tube thermal conductivity [W/(m °C)]	15
Shell (cold) side pressure drop [kPa]	50 [†]
Tube (hot) side pressure drop [kPa]	50 [†]

(†) The number of tubes and baffle spacing were selected to obtain the specified pressure drops through a numerical solving procedure.

3. Results and Discussion

3.1. Influence of Heat Exchanger Parameters

In the following, it is demonstrated how the results of a net power output comparison among pure fluids and zeotropic mixtures is affected by which heat exchanger performance parameter is used as basis for the comparison. First, the condenser size is represented using the minimum pinch point temperature difference and subsequently the condenser size is represented by mean temperature differences, $\bar{U}A$ values and heat transfer areas. The analysis presented in Section 3.1 is based on the simulation data listed in Table A1 in Appendix A.

3.1.1. Minimum Pinch Point Temperature Difference Based Comparison

Figure 3 shows the maximized net power output as a function of the condenser minimum pinch point temperature difference for i-butane, propane/i-butane (0.2/0.8), propane/i-butane (0.8/0.2) and propane. For all fluids, an approximately linear trend was observed, and the absolute difference in terms of net power output among the fluids was independent of value selected for the condenser minimum pinch point temperature difference. Sketches of the ORC unit T, s-diagrams for the four fluids are displayed in Figure 4.

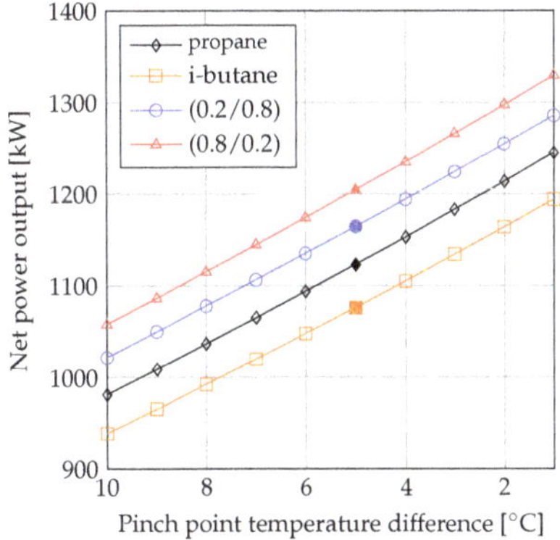

Figure 3. Net power output versus pinch point temperature difference for the condenser.

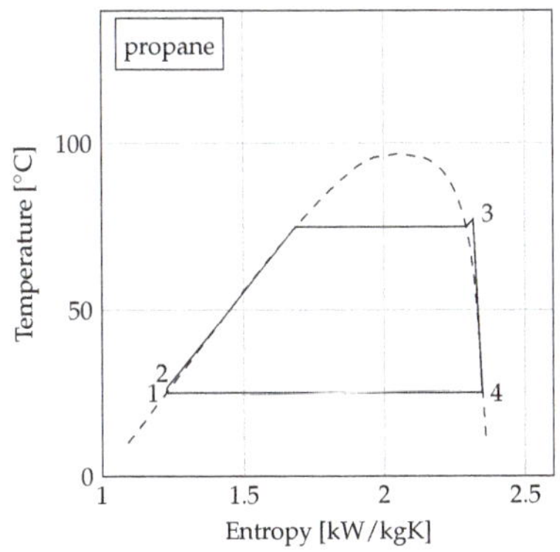

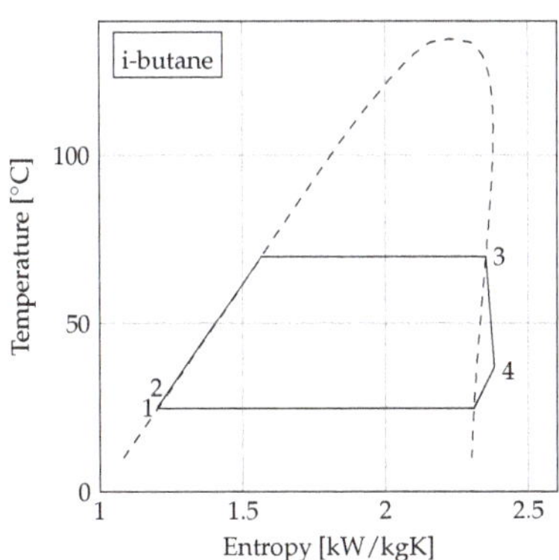

Figure 4. *Cont.*

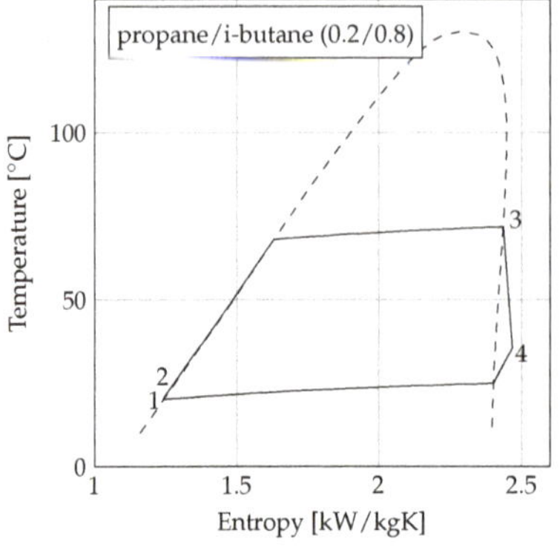
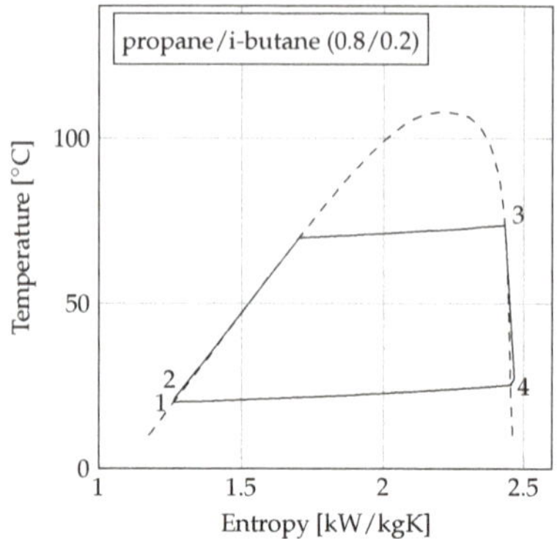

Figure 4. T, s-diagrams for the four ORC units using propane, i-butane, propane/i-butane (0.2/0.8), and propane/i-butane (0.8/0.2) with $\Delta T_{pp,cond} = 5\,°C$.

The variation of $\bar{U}A$ and ΔT_m for the condensers are depicted in Figure 5. When the minimum pinch point temperature difference was the same for the pure fluids and the mixtures, the heat transfer equipment used for mixture condensation was associated with larger values of $\bar{U}A$ and smaller values of mean temperature difference. The performance comparison was therefore affected by whether the fluids were compared based on the same values of the minimum pinch point temperature difference, mean temperature difference or $\bar{U}A$.

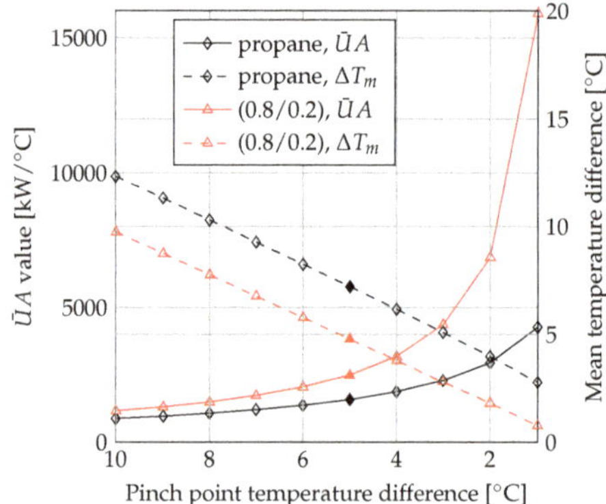

Figure 5. Condenser $\bar{U}A$ and mean temperature difference versus pinch point temperature difference.

3.1.2. Mean Temperature Difference Based Comparison

The plot depicted in Figure 6 is generated by exchanging the minimum condenser pinch point temperature difference represented on the horizontal axis in Figure 3 with the mean temperature difference. Note that the optimized solutions displayed in Figures 3 and 6 are the same. For each fluid, the left most points in Figure 6 correspond to the solutions with a minimum pinch point temperature difference in the condenser of 10 °C, while the right most points correspond to the optimized solutions with minimum pinch point temperature difference values of 1 °C. The filled marker represent the solutions with a minimum pinch point temperature difference of 5 °C.

For all fluids plotted in Figure 6, the relationship between the mean temperature difference and the net power output were approximately linear. This was also the case when the fluids were compared based on the same values of condenser pinch points. However, when compared based on the mean temperature differences, the curves for the pure fluids and the mixtures were displaced relative to each other. Generally, the mixtures achieved mean temperature differences that were similar to the pinch point temperature differences, while the pure fluids had mean temperature differences that were 1.5–3 °C higher than the pinch points. This means that the performance improvement obtained by mixtures was rated less when the comparison was based on the same values of mean temperature differences than on the same values of minimum pinch point temperature differences. The results shown in Figure 6 suggest that the performance of propane was higher than that of propane/i-butane (0.2/0.8) for mean temperature differences higher than 3 °C, whereas propane/i-butane (0.2/0.8) reached higher net power outputs than those of propane for all values of the condenser pinch point temperature difference in the pinch point based comparison (see Figure 3).

A comparison of the condensation processes for propane and propane/i-butane (0.2/0.8) is illustrated in the $\dot{Q}T$-diagrams (heat transfer rate versus temperature diagrams) depicted in Figures 7 and 8. Figure 7 displays the comparison based on equal values of minimum pinch point temperature differences ($\Delta T_{pp} = 5$ °C). The condensation process of propane is displayed as a horizontal line, since the optimum solution resulted in condensation from saturated vapor to saturated liquid. The optimum solution for propane/i-butane (0.2/0.8) involved 10 °C of desuperheating. Disregarding the desuperheating process, the condensation process for propane/i-butane (0.2/0.8) was carried out at a lower temperature than the condensation of propane, resulting in a lower mean temperature for the mixture. Due to the match between the temperature glide of the mixture and the temperature increase of the coolant, the minimum pinch point temperature difference value could be achieved throughout most of the condensation process. This resulted in dissimilar mean temperatures for the pure fluid and the mixture. The mean temperature was $\Delta T_m = 7.2$ °C for propane, while it ess $\Delta T_m = 5.2$ °C for the mixture. When the pinch point temperature difference of propane/i-butane (0.2/0.8) was increased by 2 °C, the mean temperatures of the two fluids were equal. This situation is displayed in Figure 8. When $\Delta T_m = 7.2$ °C for both fluids, the temperature difference between the coolant and the condensing working fluid was lowest for the mixture in the left half of the $\dot{Q}T$-diagram (from 0% to 50% heat transfer rate), while it was lowest for propane in the remaining part of the diagram (from 50% to 100% heat transfer rate).

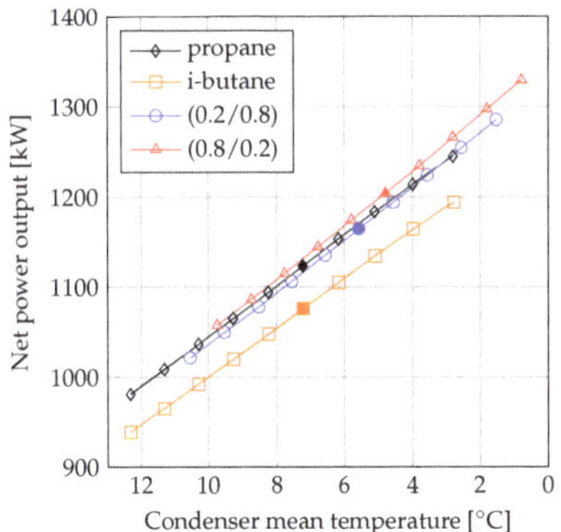

Figure 6. Net power output as a function of condenser mean temperature difference.

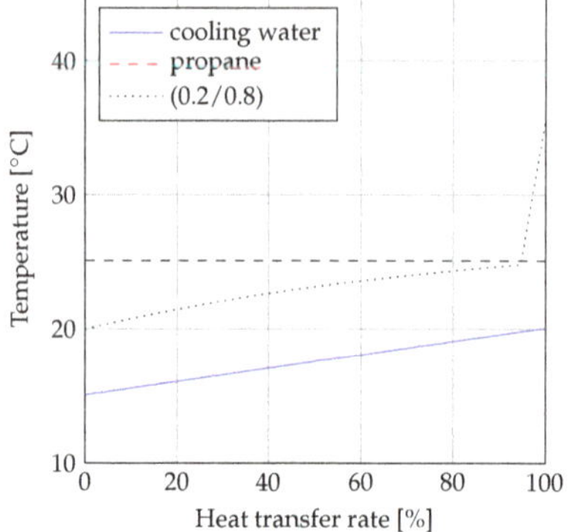

Figure 7. $\dot{Q}T$-diagram of condensation when $\Delta T_{pp} = 5\,°\mathrm{C}$ for propane and propane/i-butane (0.2/0.8).

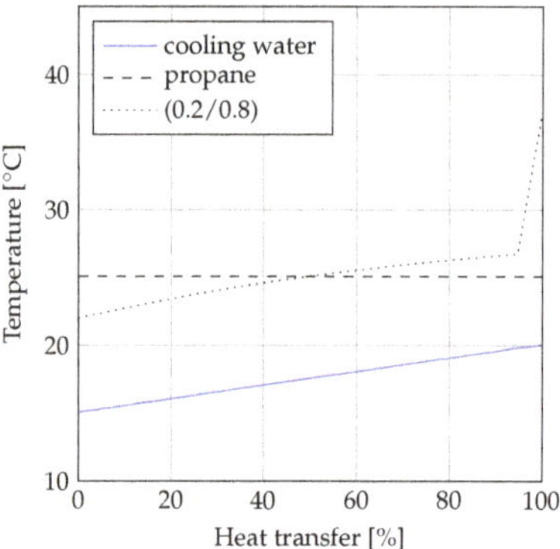

Figure 8. $\dot{Q}T$-diagram of condensation when $\Delta T_m = 7.2\,°\mathrm{C}$ for propane and propane/i-butane (0.2/0.8).

3.1.3. $\bar{U}A$ Value Based Comparison

The plot depicted in Figure 9 displays the comparison of propane, i-butane, propane/i-butane (0.2/0.8), and propane/i-butane (0.8/0.2). Figure 9 is an alternative representation of the results presented in Figures 3 and 6. Essentially, the difference between Figures 3 and 9 is that the net power outputs are plotted as a function of $\bar{U}A$ instead of minimum pinch point temperature differences. The left most values (lowest $\bar{U}A$ and net power output) plotted in Figure 9 thereby correspond to the solutions with minimum pinch point temperature difference values of 10 °C, while the right most values (highest $\bar{U}A$ and net power output) correspond to the solutions with minimum pinch point temperature values of 1 °C. Note that the solutions with 1 °C minimum pinch point temperature difference for the mixtures are outside the boundaries of the plot. The filled markers correspond to the solutions with minimum pinch point temperature differences of 5 °C.

In the comparison between propane and propane/i-butane (0.2/0.8), propane achieved the highest performance for $\bar{U}A$ values below 4000 kW/°C. The plots in Figure 9 suggest that the benefits of using mixtures were highest when the value of $\bar{U}A$ for the condenser was high. Furthermore, by comparing the net power outputs of the fluids based on the same values of the condenser $\bar{U}A$, the increase in net power output was lower compared to the case when the net power outputs were compared based on the same minimum pinch point temperature differences. For example, the net power output of the mixture propane/i-butane (0.8/0.2) was 7.0% higher compared to propane for condenser minimum pinch point temperature differences of 3 °C. Simultaneously, the $\bar{U}A$ value of the condenser was 91% higher for the mixture compared to propane. When the net power outputs were compared based on condenser $\bar{U}A$ values of 4300 kW/°C (corresponding to $\Delta T_{pp,cond} = 3$ °C for the mixture and $\Delta T_{pp,cond} = 1$ °C for propane) the net power output of the mixture was 1.7% larger compared to that of propane. Comparing the fluids based on the same $\bar{U}A$ values was similar to the comparison based on equal mean temperature differences, since the variation in heat transfer rate was insignificant. These results indicate that the comparison based on the same minimum pinch point temperature differences resulted in an optimistic estimation of the thermodynamic benefit of using zeotropic mixtures, while the comparisons based on the same $\bar{U}A$ values and mean temperature difference resulted in conservative estimations of the benefits of using zeotropic mixtures.

3.1.4. Heat Transfer Area Based Comparison

Figure 10 shows the results of the fluid comparison when the fluids are rated on equal values of heat transfer area for the condenser. Similar to Figure 9, the plots in Figure 10 were generated based on the optimized solutions from Figure 3. The heat transfer areas were calculated based on the outputs (pressures, temperatures and mass flow rates) from the thermodynamic ORC process model by using the shell-and-tube heat exchanger model with conditions listed in Table 3. For each fluid, the lowest condenser areas in Figure 10 correspond to minimum pinch point temperature difference values of 10 °C, while the highest condenser areas correspond to minimum pinch point temperature values of 1 °C. The filled markers denote the solutions with a minimum pinch point temperature difference of 5 °C.

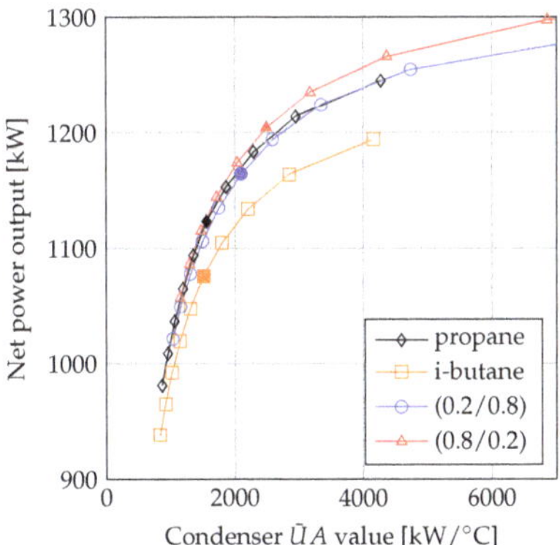

Figure 9. Net power output as a function of condenser $\bar{U}A$.

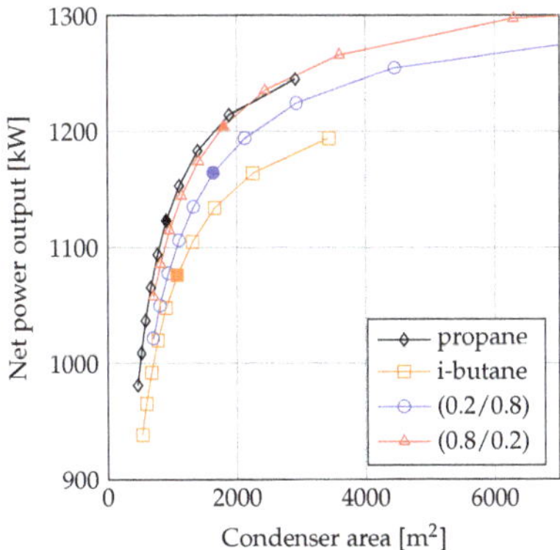

Figure 10. Net power output as a function of condenser heat transfer area.

By comparing Figures 9 and 10, the influence of the variation in $\bar{U}$ can be observed. The heat transfer performance of propane was higher than that of i-butane. Furthermore, the overall heat transfer coefficient of propane was also larger than those of the mixtures. In the comparisons depicted in Figure 10, the larger heat transfer coefficient of propane increased the feasibility of this fluid compared to the $\bar{U}A$ value based method presented in Figure 9.

At condenser areas below 2000 m^2, propane was the highest performing fluid. At around 2000 m^2, there was a crossing of the curves for propane and propane/i-butane (0.8/0.2), and at higher condenser areas the (0.8/0.2) mixture achieved the highest net power outputs. The curve crossing occurred where the minimum pinch point temperature difference in the condensers was around 2 °C for propane and 5 °C for the mixture.

3.2. Fluid Performance Ranking

In this section, it is demonstrated how the outcome of a fluid performance ranking can be affected by the employed modeling method. The demonstration is based on the fluid selection study presented by Andreasen et al. [10] for a 120 °C liquid water heat source, and the fluids considered for comparison are the 30 optimum fluids identified in that study. In the present study, the net power output of the ORC unit was maximized for each of the 30 fluids by employing the modeling conditions listed in Table 1. Supercritical pressures were allowed and the maximum pressure was limited to 100 bar. The high pressure in the system was limited to be below 80% of the critical pressure for subcritical systems and above 120% of the critical pressure for transcritical systems. For transcritical ORC solutions, the degree of superheating was replaced by the temperature difference between the expander inlet temperature and the critical temperature as decision variable which was varied in the range from 0 °C to 100 °C. For mixtures, the composition was optimized.

The ranking of the 30 working fluids based on fixed minimum pinch point temperature differences of $\Delta T_{pp,PrHE} = 10$ °C and $\Delta T_{pp,cond} = 5$ °C is listed in Table 4. The ranking was essentially the same as that presented by Andreasen et al. [10] except from the definition of the expander efficiency, which Andreasen et al. [10] based on a polytropic efficiency of 0.8, while an isentropic efficiency of 0.8 was employed in the present study. Moreover, the pressure constraints for subcritical and transcritical ORC solutions were not employed by Andreasen et al. [10]. This resulted in small variations in the fluid ranking compared to the previous results.

Table 5 lists the ranking of the 30 fluids when based on a fixed value of $\bar{U}A_{tot} = 3500$ kW/°C. This value is the sum of the primary heat exchanger and condenser $\bar{U}A$ values for R218 in Table 4. The distribution of the total $\bar{U}A$ value between the primary heat exchanger and condenser was included as a decision variable. The optimization of the $\bar{U}A$ distribution enabled the power output of the ORC unit with R218 to increase from 1444 kW to 1460 kW. Despite this increase, the fluid was not ranked as the best performing fluid when the same value of $\bar{U}A_{tot}$ was assumed for all fluids. Instead, the results suggest that R143a was the highest performing fluid due to an increase in net power output from 1361 kW to 1521 kW. This large increase was due to the relatively low value of $\bar{U}A_{tot}$ resulting from the minimum pinch point temperature difference based optimization (see Table 4). When comparing the net power output and $\bar{U}A$ value of R218 and R143a in Table 4, R218 reached 6% higher net power output with 22% higher $\bar{U}A_{tot}$. The minimum pinch point temperature difference based method resulted in large differences in the requirement for the heat exchangers. When rated based on the $\bar{U}A$ value method, the differences due to variations in the $\bar{U}A$ values were eliminated, which enabled a fairer comparison of the fluids. The fluids R227ea, R1234yf and propylene also experienced a large increase in the ranking when rated on the $\bar{U}A$ value based method rather than the minimum pinch point temperature difference based method.

Generally, the results indicate that, when the fluid ranking was based on equal values of $\bar{U}A_{tot}$, the pure fluids performed better relative to the zeotropic mixtures than when the performance comparison was based on equal values of the minimum pinch point temperature difference. In the minimum pinch point temperature difference based comparison, 19 out of the 30 fluids were zeotropic mixtures (including the predefined mixture R422A). In this case, the zeotropic mixtures enabled up to 13.6% higher performance compared to the best pure fluid in each mixture. In the $\bar{U}A_{tot}$ based comparison, only 12 fluids were zeotropic mixtures, since the composition of seven mixtures converged to 0% or 100% mole fraction. In this case, the performance increase of the zeotropic mixtures compared to the best pure fluid in each mixture was below 2.6%. High performance was reached for the transcritical mixture ethane/propane in both fluid rankings, but the composition changed from 88.3% (in Table 4) to 3.73% (in Table 5). The mixtures propane/butane, propane/i-butane, propane/pentane, propane/i-pentane and propane/hexane did not reach higher net power output than pure propane. It is also worth noting that, although ethane as a pure fluid reached the lowest net power output of the 30 fluids, it was possible to reach higher net power outputs than propane, butane, i-butane, pentane, and i-pentane by mixing small amounts of ethane in these fluids. The propane/i-butane mixture, which was analyzed thoroughly (see Section 3.1), did not reach higher performance than pure propane.

Table 4. Fluid selection and optimization based on fixed values of the minimum pinch point temperature difference $\Delta T_{pp,PrHE} = 10\,^{\circ}\text{C}$ and $\Delta T_{pp,cond} = 5\,^{\circ}\text{C}$.

Fluid	$\dot{W}_{net}$ (kW)	X_{wf} (%)	P_{PrHE} (bar)	ΔT [‡] (°C)	$\dot{m}_{wf}$ (kg/s)	T_{cond} (°C)	$\bar{U}A_{cond}$ (kW/°C)	$\bar{U}A_{PrHE}$ (kW/°C)	ΔT_g [‡‡] (°C)	$\Delta \dot{W}_r$ [‡‡‡] (%)
R218[tc]	1444	-	43.1	28.5	163	24.0	2075	1426	-	-
R422A[tc]	1431	-	52.8	27.3	109	23.4	2173	1182	1.36	1.75
R125[tc]	1406	-	55.4	33.4	118	24.6	1991	1203	-	-
R41[tc]	1367	-	98.7	65.9	48.3	24.2	1881	1182	-	-
eth./pro.[tc]	1367	88.3	87.1	66.3	44.4	20.0	2288	1354	5.90	12.6
R143a[tc]	1361	-	50.8	25.8	82.5	25.0	1823	1046	-	-
eth./ibut.[tc]	1334	95.5	90.4	68.7	45.5	20.2	2197	1384	7.63	9.90
eth./but.[tc]	1330	96.3	91.5	68.9	45.5	20.3	2183	1384	8.16	9.53
eth./ipen.[tc]	1316	97.7	94.1	69.6	46.2	20.4	2174	1391	9.09	8.37
eth./pen.[tc]	1310	97.9	92.9	68.4	45.7	20.4	2116	1385	13.1	7.88
SF6[tc]	1296	-	78.3	62.5	156	23.4	1814	1404	-	-
eth./hex.[tc]	1291	99.1	95.1	72.0	46.2	20.9	2095	1371	13.1	6.31
eth./hep.[tc]	1285	99.2	94.6	71.3	46.1	21.0	2068	1370	25.7	5.82
CO2/pro.[tc]	1233	74.5	100	68.7	65.1	20.0	1901	1266	11.9	9.76
R227ea[sc]	1215	-	17.0	0	99.0	24.5	1710	688.5	-	-
R1234yf[sc]	1215	-	25	0.685	79.8	24.9	1676	689.7	-	-
ethane[tc]	1214	-	94.4	77.8	46.3	23.0	1737	1304	-	-
CO2/ibut.[tc]	1211	88.5	100	65.2	65.7	20.0	1923	1202	16.0	12.5
C4F10[sc]	1205	-	10.1	0	121	24.0	1751	739.7	-	-
CO2/but.[tc]	1199	91.0	100	66.7	66.0	20.0	2012	1165	15.6	13.6
pro./ibut.[sc]	1192	79.6	22.2	0	33.5	20.4	2286	591.3	5.26	6.14
pro./but.[sc]	1186	91.5	24.6	0.730	33.1	21.1	2193	591.3	4.96	5.62
pro./ipen.[sc]	1179	96.8	25.8	1.79	33.2	21.7	2077	592.0	5.66	4.98
R32[sc]	1178	-	46.3	42.5	38.1	24.5	1558	627.0	-	-
pro./pen.[sc]	1177	96.6	25.2	2.17	32.8	21.3	2058	590.4	8.20	4.77
pro./hex.[sc]	1165	99.0	26.7	4.02	32.1	22.3	1922	581.3	8.42	3.75
ibut./ipen.[sc]	1150	86.6	9.29	0	31.0	20.5	2048	539.5	6.15	6.84
but./ipen.[sc]	1142	72.0	6.25	0	28.5	20.2	2081	523.4	5.39	8.18
but./pen.[sc]	1136	83.4	6.62	0	27.9	20.3	1992	521.5	6.52	7.60
propylen[sc]	1125	-	34.1	7.12	33.5	25.0	1555	599.1	-	-

(‡) ΔT represents the degree of superheating for subcritical ORC units (sc) and the temperature difference between the turbine inlet and the critical temperature for transcritical ORC units (tc). (‡‡) ΔT_g represents the temperature glide of condensation. (‡‡‡)] $\Delta \dot{W}_r$ represents the relative difference in net power output between the mixture (mix) and the best pure component in the mixture (pure), $\Delta \dot{W}_r = (\dot{W}_{net,mix} - \dot{W}_{net,pure})/\dot{W}_{net,pure}$.

Moreover, when the fluid ranking was based on equal values of the minimum pinch point temperature difference, 17 out of the 30 fluids were transcritical, but, in the fluid ranking based on equal values of $\bar{U}A_{tot}$, only seven fluids were transcritical. The highest net power output was still achieved with transcritical ORC units, however the relative difference in net power output of the highest performing subcritical compared to the highest performing transcritical unit was reduced from 18.8% (R227ea compared to R218 in Table 4) to 1.1% (R227ea compared to R143a in Table 5). This indicates that the method assuming the same minimum pinch point temperature differences for all fluids resulted in optimistic performance estimations of the thermodynamic benefits of using transcritical ORC units and/or zeotropic mixtures, while the method assuming the same total $\bar{U}A$ value resulted in a conservative estimations.

Table 5. Fluid selection and optimization based on fixed values of $\bar{U}A_{tot} = \bar{U}A_{PrHE} + \bar{U}A_{cond} = 3500$ kW/°C.

Fluid	$\dot{W}_{net}$ (kW)	X_{wf} (%)	P_{PrHE} (bar)	ΔT [‡] (°C)	$\dot{m}_{wf}$ (kg/s)	T_{cond} (°C)	$\bar{U}A_{cond}$ (kW/°C)	$\bar{U}A_{PrHE}$ (kW/°C)	ΔT_g [‡‡] (°C)	$\Delta\dot{W}_r$ [‡‡‡] (%)
R143a[tc]	1521	-	47.9	36	72.4	24.4	1819	1681	-	-
R422a[tc]	1511	-	50.7	35	96.5	23.8	1831	1669	1.35	1.00
R227ea[sc]	1505	-	23.3	0.14	105	24.4	1870	1630	-	-
R125[tc]	1496	-	54.0	42	104	24.3	1850	1650	-	-
R1234yf[sc]	1475	-	27.1	1.3	92.9	25.0	1947	1551	-	-
R218[tc]	1460	-	41.8	32	149	24.3	1843	1657	-	-
R41[tc]	1457	-	96.3	72	47.6	24.3	1857	1643	-	-
C4F10[sc]	1455	-	13.3	0	121	23.6	1877	1623	-	-
CO2/pro.[sc]	1442	0.747	34.2	4.8	38.0	23.6	1800	1699	1.76	1.24
eth./pro.[sc]	1441	3.73	34.6	5.5	38.0	22.8	1950	1550	2.44	1.20
R32[sc]	1439	-	46.3	50	43.2	23.6	2065	1435	-	-
propylen[sc]	1425	-	36.4	26	32.3	23.3	2021	1479	-	-
pro./ibut.[sc]	1424	100	34.0	3.8	37.6	24.4	1951	1549	0	0
pro./but.[sc]	1424	100	34.0	3.9	37.6	24.4	1951	1549	0	0
pro./ipen.[sc]	1424	100	34.0	3.8	37.6	24.4	1950	1550	0	0
pro./pen.[sc]	1424	100	34.0	3.8	37.6	24.4	1952	1548	0	0
pro./hex.[sc]	1424	100	33.9	3.8	37.6	24.4	1952	1548	0	0
eth./ibut.[sc]	1407	3.68	14.3	0	33.8	18.0	2092	1408	7.14	2.25
CO2/ibut.[sc]	1400	1.19	13.8	0	33.9	17.7	2087	1415	6.77	1.76
ibut./ipen.[sc]	1391	92.0	12.1	0	32.5	21.4	2134	1358	4.00	1.08
eth./but.[sc]	1389	2.27	10.3	0	30.1	17.2	2115	1385	7.49	2.56
but./ipen.[sc]	1387	51.1	6.39	0.23	30.3	19.6	2125	1374	6.56	2.41
but./pen.[sc]	1382	82.6	7.86	0.23	29.6	20.1	2202	1271	6.73	2.00
CO2/but.[sc]	1373	0.406	9.79	0.19	29.9	19.5	2179	1321	3.94	1.34
eth./ipen.[sc]	1365	0.412	4.30	0	30.6	17.2	2142	1358	6.43	1.50
eth./pen.[sc]	1362	0.541	3.43	0	29.0	17.1	2132	1368	6.67	1.75
SF6[tc]	1336	-	75.0	64	155	23.3	1857	1643	-	-
eth./hex.[sc]	1331	0	1.23	0	28.3	22.6	2235	1265	0	0
eth./hep.[sc]	1327	0	0.479	0	28.1	22.5	2241	1259	0	0
ethane[tc]	1281	-	94.7	81	45.7	22.5	1899	1601	-	-

(‡) ΔT represents the degree of superheating for subcritical ORC units (sc) and the temperature difference between the turbine inlet and the critical temperature for transcritical ORC units (tc). (‡‡) ΔT_g represents the temperature glide of condensation. (‡‡‡) $\Delta\dot{W}_r$ represents the relative difference in net power output between the mixture (mix) and the best pure component in the mixture (pure), $\Delta\dot{W}_r = (\dot{W}_{net,mix} - \dot{W}_{net,pure})/\dot{W}_{net,pure}$.

3.3. Assessment of Modeling Methods

The use of zeotropic mixtures provides the flexibility of adjusting the temperature profile of evaporation and condensation to obtain optimal alignment with the temperature profiles of sensible heat sources and sinks. This has an effect on the temperature difference between the heat exchanging fluid streams throughout the primary heat exchanger and the condenser. By basing the comparison of pure fluids and mixtures on equal values of the minimum pinch point temperature difference, there is no distinguishing between solutions where the minimum temperature difference value is reached at a single location in the heat exchanger (for example in case of pure fluids) and solutions where the temperature difference is close to the minimum value throughout the heat exchanger (in case of temperature profile matching) (see Figure 7). However, as demonstrated in Section 3.1, these two situations can result in very different mean temperatures and $\bar{U}A$ values. Ultimately, when comparing pure fluids and zeotropic mixtures based on equal values of minimum pinch point temperature difference, the zeotropic mixtures benefit from reduced irreversibilities in the heat transfer processes due to temperature profile matching. However, the additional cost related to the larger heat transfer equipment required for transferring the heat across a lower mean temperature difference is unaccounted for. Thereby, the estimation of the benefit of temperature profile matching is deemed to be optimistic when the comparison is based on equal values of minimum pinch point temperature differences, since the ORC unit employing the zeotropic mixtures are provided with the additional benefit of having heat exchangers with larger thermal capacity (larger $\bar{U}A$ and lower mean temperature difference). Using this thermodynamic modeling approach for comparing the performance of pure fluids and zeotropic mixtures there is a risk of overestimating the benefits of zeotropic mixtures, which might not be representative of the actual performance difference when the cost of heat exchangers are taken into account in subsequent thermoeconomic analyses.

By use of alternative modeling methods, it is possible, with limited additional modeling effort, to take differences in mean temperature differences or $\bar{U}A$ values into account. The added modeling complexity solely consists of discretizing the heat exchangers in a suitable number of control volumes, which enables to calculate the $\bar{U}A$ values and mean temperature differences, and defining limiting values for either of them. In many situations it is even required to discretize the heat transfer process for identifying the location of the minimum pinch point temperature difference, for example when the temperature profile of the mixture is highly curved, or when the high pressure is supercritical or slightly below the critical pressure. It is beneficial to specify limits for the mean temperature difference or the $\bar{U}A$ value of the heat exchangers rather than for the minimum pinch point temperature difference, since the mean temperature difference and the $\bar{U}A$ value take the complete heat transfer process into account and are therefore more indicative of the heat transfer surface area and thereby the cost of the heat exchanger. The use of the methods assuming either the same mean temperature difference or the same $\bar{U}A$ values for pure fluids and zeotropic mixtures, is deemed to result in conservative estimations of the thermodynamic performance benefit of using zeotropic mixtures.

Defining a constraint based on the mean temperature difference is beneficial in the sense that it is conceptually similar to the minimum pinch point temperature difference and it is therefore easier to define suitable limiting values. Defining heat exchanger constraints based on $\bar{U}A$ values provides an additional option of optimizing the distribution of $\bar{U}A$ to each of the heat exchangers by defining a limit for the total ORC unit $\bar{U}A$ (sum of $\bar{U}A$ for all heat exchangers in the unit). The drawback of defining constraints based on $\bar{U}A$ values is that the $\bar{U}A$ value depends on the capacity of the system, i.e., the net power output and heat transferred in the process. One way of addressing this issue is to define a base case fluid for which the heat exchanger minimum pinch point temperature differences are fixed at reasonable values. The $\bar{U}A$ values of the heat exchangers for the base case solution can then be used as limiting values for the remaining simulations. The use of methods fixing the mean temperature differences or $\bar{U}A$ values of heat exchangers does not account for differences in heat transfer coefficients. Such differences can be accounted for by including heat exchanger models in the optimization framework for estimation of heat transfer areas. However, this requires the use of heat transfer correlations which adds to the model complexity and computational time required for the simulations. In relation to fluid screening studies, it is desirable to limit the computational time of the model when many fluids are evaluated. An overview of the method assessment is shown in Table 6.

The thermodynamic modeling methods discussed in this paper do not account for effects of fluid properties on the design, efficiency and cost of the ORC expander. Such aspects are essential when selecting a fluid for an ORC unit. Astolfi et al. [45], Martelli et al. [46], and Meroni et al. [47] provided methods for including axial turbine design aspects in the framework of thermodynamic and thermoeconomic ORC process optimizations. Such methods are relevant to combine with the methods discussed in the present paper for preliminary design and optimization of ORC systems.

Table 6. Method assessment overview.

	Benefits	Drawbacks
Fixed ΔT_{pp}	• Simple model allowing quick calculations • Easy to define reasonable minimum pinch point temperature difference values independently of the ORC unit capacity	• Mixtures are systematically allowed to have lower mean temperature differences than pure fluids • Influence of temperature profile matching on heat exchanger size is not accounted for • Differences in heat transfer coefficients are not accounted for
Fixed ΔT_m	• Influence of temperature profile matching on heat exchanger size is accounted for • Same mean temperatures for all fluids • Easy to define reasonable mean temperature difference values independently of the ORC unit capacity	• Higher model complexity and computational time compared to the fixed minimum pinch point temperature difference method • Differences in heat transfer coefficients are not accounted for
Fixed UA	• Influence of temperature profile matching on heat exchanger size is accounted for • Same $\bar{U}A$ values for all fluids • Possible to select total ORC unit $\bar{U}A$ value and optimize distribution to each heat exchanger • Conceptually similar to fixing the heat transfer area in case of equal values of $\bar{U}$	• Higher model complexity and computational time compared to the fixed minimum pinch point temperature difference method • Value of $\bar{U}A$ must be selected depending on the ORC unit capacity • Differences in heat transfer coefficients are not accounted for
Fixed A	• Influence of temperature profile matching on heat exchanger size is accounted for • Same heat transfer area for all fluids • Differences in heat transfer coefficients are accounted for	• Higher model complexity and computational time compared to the three other methods • Value of heat transfer area, A, must be selected depending on the ORC unit capacity

3.4. Feasibility of Using Zeotropic Mixtures

Traditionally, the primary reason for using zeotropic mixtures as working fluids in ORC power systems is to utilize the temperature glide of phase change for temperature profile matching with sensible heat sources and sinks, with the aim of reducing the irreversibilities of the heat transfer processes. As indicated in the fluid comparison based on equal heat transfer area (see Figure 10), those benefits are highest when the heat transfer area is high. In the theoretical case of infinitely large counter-flow heat exchangers, it is possible for mixtures to reach higher capacities (lower mean temperature differences and higher $\bar{U}A$) than pure fluids, since the temperature profiles achieve a better match with the sensible heat source and sink in the primary heat exchanger and the condenser. On the other hand, when the heat exchangers are small, the temperature differences in the heat exchangers are large and the $\bar{U}A$ values are small. In this case, the performance increase due to the better temperature match achieved with zeotropic mixtures is limited and tends to be overcompensated by the degradation in heat transfer coefficient, as demonstrated in Section 3.1.4. This observation indicates that zeotropic mixtures are more likely to be economically feasible compared to pure fluids in applications where it pays off to invest in large heat exchangers. This is the case in geothermal applications, where a large investment is required for establishing the geothermal well. Astolfi et al. [45] demonstrated how their thermoeconomic optimization results for pure fluids in an ORC unit designed for utilizing geothermal heat, was affected by the investment required for the geothermal well. A variation of the geothermal well cost from zero to 12 M€ resulted in an increase of the ORC unit cost (power block only) from 3.03 M€ to 8.57 M€, and a reduction of the minimum pinch point temperature difference of the condenser from 2.5 °C to 0.43 °C. Similarly, it is more likely that zeotropic mixtures achieve higher economic performance than pure fluids, when the cost of the heat exchangers is low.

It is also worthwhile considering zeotropic mixtures for other reasons than the temperature profile matching capabilities. By including zeotropic mixtures in the fluid screening, the number of considered working fluids is larger, and it is more likely to identify fluids with desirable properties considering multiple criteria such as thermodynamic performance, environmental friendliness, safety and cost. For example, the high flammability of hydrocarbons could be reduced by adding a non-flammable fluid [48,49]. Fluids which result in high process pressures could benefit from being mixed with a low pressure fluid. This could result in a reduction in piping and heat exchanger costs [25]. Additionally, for the same condenser bubble point (condensing fluid outlet temperature), it is possible for zeotropic mixtures to have a higher increase in the coolant temperature due to the temperature glide. For the same amount of heat transfer, the mass flow rate of cooling water can thereby be smaller for mixtures. This can result in lower size and cost for cooling water circulating pumps, and cooling towers [5].

3.5. Method Selection for Other Thermodynamic Processes

Temperature profile matching is not only utilized for zeotropic mixtures, but also for pure fluids where temperature profile matching in the preheater has been used for optimizing the ORC system performance [50,51]. In addition, the use of a recuperator can result in almost perfect alignment of the temperature profiles of the heated liquid and the cooled vapor [51]. In such situations, it is important to check that the mean temperature differences and the $\bar{U}A$ values do not get excessively low and high, respectively. In power cycle units where the pressure is supercritical, for example transcritical ORC units or supercritical CO_2 units, temperature profile matching is possible in the heat transfer process between the hot fluid and the supercritical working fluid. When comparing such units with power cycle units using subcritical pressures it is also important to account for differences in mean temperature differences and $\bar{U}A$ values in order not to overestimate the performance of the units employing supercritical pressures (see also Section 3.2). Due to the curved temperature profile of the supercritical heating process it is necessary to discretize the heat exchanger for identifying the pinch point location. Thus, it is possible to define constraints on the mean temperature difference or $\bar{U}A$ values of the heat exchangers with limited additional effort compared to applying minimum pinch point temperature difference constraints.

The implementation of constraints on either mean temperature differences or $\bar{U}A$ values rather than minimum pinch point temperature differences, should also be considered for other power cycles utilizing zeotropic mixtures as working fluids. Examples of such cycles are the Kalina cycle using ammonia/water mixtures, and physical absorption cycles using lithium-bromide/water mixtures.

The alternative methods discussed in this paper are also useful in combination with computer aided molecular design (CAMD) [52–54], and continuous molecular targeting (CoMT) [55,56] methods, which consider the molecular optimization of the working fluid. In such optimization problems, the ORC process model is typically based on thermodynamics, since a full thermoeconomic model including heat exchanger sizing would be too extensive to implement at this stage. Although the CAMD and CoMT methods are able to identify thermodynamic optima, it is possible that the fluids designed are not thermoeconomically optimal. By using minimum pinch point temperature difference based models, there is a risk that benefits due to temperature profile matching is overestimated. It is therefore important to check whether there is a correlation between $\bar{U}A$ values or mean temperatures, with the performance ranking of the designed fluids. If this is the case, it is worthwhile considering implementation of constraints on mean temperature differences or $\bar{U}A$ values. Cignitti et al. [57] carried out a CAMD based optimization integrating ORC process and working fluid design, and found different optimum working fluids depending on whether the selected heat exchanger constraints were based on minimum pinch point temperature differences or $\bar{U}A$ values.

4. Conclusions

In this paper, four different methods for performance comparison of pure fluids and zeotropic mixtures in ORC systems were assessed. The methods were characterized by which modeling approach was used for representing the performance of the heat exchangers in the systems. The methods compared the net power output of the fluids based on either the same minimum pinch point temperature differences, the same mean temperature differences, the same $\bar{U}A$ values, or the same heat transfer areas for all fluids.

The comparison of propane, i-butane, propane/i-butane (0.2/0.8) and propane/i-butane (0.8/0.2) based on the same values of condenser pinch point temperature differences suggests that the highest net power outputs are achieved by the mixtures for all values of condenser pinch temperature difference. When compared based on the same values of condenser mean temperature differences or $\bar{U}A$ values, the results suggest that propane/i-butane (0.8/0.2) achieves the highest net power output. However, the mixture propane/i-butane (0.2/0.8) is outperformed by propane at condenser mean temperature differences above 3 °C and $\bar{U}A$ values below 4000 kW/°C. When comparing the net power outputs of the fluids based on the same condenser heat transfer areas, the highest performance is achieved by propane at heat transfer areas below 2000 m^2, while propane/i-butane (0.8/0.2) achieves the highest performance at heat transfer areas above 2000 m^2. In the heat transfer area based comparison, propane achieves higher net power outputs than propane/i-butane (0.2/0.8) for all considered values of the condenser heat transfer area. The results suggest that the fluid i-butane achieves the lowest net power outputs of the four fluids in all the considered cases.

A fluid performance ranking of 30 working fluids assuming the same minimum pinch point temperature differences for all fluids, indicates that the net power outputs of zeotropic mixtures are up to 13.6% higher compared to the best pure fluids in the mixtures. On the other hand, by assuming that the sum of the condenser and primary heat exchanger $\bar{U}A$ value is the same for all fluids, the results indicate that the net power outputs of zeotropic mixtures are only up to 2.56% higher compared to the best pure fluids in the mixtures. Similarly, the estimated performance benefit of using transcritical ORC units decreases from 18.8% in the minimum pinch point temperature difference based comparison to 1.1% in the $\bar{U}A$ value based comparison.

The method assuming equal minimum pinch point temperature differences for all fluids was deemed to result in optimistic estimations of the benefits of using zeotropic mixtures, while the methods assuming equal mean temperature differences or $\bar{U}A$ values were deemed to result in conservative estimations. The method assuming equal heat transfer areas for all fluids was deemed too comprehensive for thermodynamic optimization considering modeling complexity and computational time. This indicates that it is beneficial to use the method assuming equal minimum pinch point temperature differences and the method assuming equal $\bar{U}A$ values (or mean temperature differences) concurrently in preliminary fluid selection, since they result in different conclusions regarding the thermodynamic performance of working fluids.

Relevant future work within this research topic could comprise an economic optimization of an ORC system considering a group of selected working fluids. By carrying out the working fluid selection for the economic analysis based on two or more of the thermodynamic methods assessed in the present paper, it is possible to judge which of the methods identifies the more economically beneficial working fluids.

Author Contributions: Conceptualization, J.G.A., M.R.K. and F.H.; Formal analysis, J.G.A.; Funding acquisition, F.H.; Investigation, J.G.A.; Methodology, J.G.A., M.R.K. and F.H.; Project administration, F.H.; Supervision, M.R.K. and F.H.; Validation, J.G.A. and M.R.K.; Writing—original draft, J.G.A.; and Writing—review and editing, M.R.K. and F.H.

Funding: This research was funded by Innovationsfonden, The Danish Council for Strategic Research in Sustainable Energy and Environment project ID: 1305-00036B.

Acknowledgments: The work presented in this paper was conducted within the frame of the THERMCYC project ("Advanced thermodynamic cycles utilising low-temperature heat sources", project ID: 1305-00036B;

see www.thermcyc.mek.dtu.dk) funded by Innovationsfonden, The Danish Council for Strategic Research in Sustainable Energy and Environment. The financial support is gratefully acknowledged.

Conflicts of Interest: The authors declare no conflict of interest.

Nomenclature

Abbreviations

CAMD	Computer aided molecular design
CoMT	Continuous molecular targeting
ORC	Organic Rankine cycle

Greek symbols

α	Convective heat transfer coefficient, (kW/m^2 °C)
Δ	Difference
η	Efficiency, (%)
λ	Thermal conductivity, (kW/m°C)

Symbols

A	Area, (m^2)
d	Diameter, (m)
h	Specific enthalpy, (kJ/kg)
L	Tube length, (m)
l_{bc}	Length of baffle cut, (m)
$\dot{m}$	Mass flow rate, (kg/s)
n	Number of control volumes, (-)
N_t	Number of tubes, (-)
P	Pressure, (bar)
p_t	Tube pitch, (m)
$\dot{Q}$	Heat transfer rate, (kW)
T	Temperature, (°C)
$\bar{U}$	Average overall heat transfer coefficient, (kW/m^2 °C)
$\dot{W}$	Mechanical power, (kW)
X	Mole fraction, (-)

Subscripts and superscripts

c	Cold
cond	Condenser
crit	Critical
g	Glide of condensation
hf	Hot fluid
i	Input
in	Inner
j	Control volume counter
lm	Log mean
m	Mean
mix	Mixture
net	Net
o	Out
ou	Outer
pp	Pinch point
PrHE	Primary heat exchanger
pure	Pure fluid
r	Relative
sc	Subcritical
tc	Transcritical
tot	Total
wf	Working fluid

Appendix A

Table A1 displays the variation in net power output, condenser mean temperature difference, condenser $\bar{U}A$ values, and condenser areas as functions of the condenser pinch point temperature difference for the working fluids propane, i-butane, propane/i-butane (0.2/0.8) and propane/i-butane (0.8/0.2). The data are plotted in Figures 3, 5, 6, 9, and 10.

Table A1. Variation of net power output ($\dot{W}_{net}$), condenser mean temperature difference ($\Delta T_{m,cond}$), condenser $\bar{U}A$ values ($\bar{U}A_{cond}$), and condenser areas (A_{cond}) as functions of the condenser pinch point temperature difference ($\Delta T_{pp,cond}$) for propane, i-butane and the mixtures propane/i-butane (0.2/0.8) and propane/i-butane (0.8/0.2).

$\Delta T_{pp,cond}$ [°C]		10	9	8	7	6	5	4	3	2	1
$\dot{W}_{net}$ [kW]	propane	981	1009	1037	1065	1094	1123	1153	1183	1214	1245
	i-butane	939	965	992	1020	1048	1076	1105	1134	1164	1194
	(0.2/0.8)	1022	1050	1078	1107	1135	1165	1194	1224	1255	1286
	(0.8/0.2)	1058	1086	1115	1144	1174	1204	1235	1266	1298	1330
$\Delta T_{m,cond}$ [°C]	propane	12.3	11.3	10.3	9.3	8.2	7.2	6.2	5.1	4.0	2.8
	i-butane	12.3	11.3	10.3	9.3	8.2	7.2	6.2	5.1	4.0	2.8
	(0.2/0.8)	10.6	9.5	8.5	7.5	6.6	5.6	4.6	3.6	2.5	1.5
	(0.8/0.2)	9.8	8.8	7.8	6.8	5.8	4.8	3.8	2.8	1.8	0.8
$\bar{U}A_{cond}$ [kW/°C]	propane	865	957	1065	1200	1362	1572	1865	2286	2953	4273
	i-butane	835	924	1025	1154	1315	1527	1808	2207	2858	4164
	(0.2/0.8)	1041	1163	1319	1511	1761	2100	2590	3357	4740	8008
	(0.8/0.2)	1155	1302	1486	1726	2041	2492	3185	4370	6859	15925
A_{cond} [m²]	propane	454	508	575	660	765	905	1105	1402	1892	2909
	i-butane	527	591	672	773	902	1078	1317	1664	2249	3439
	(0.2/0.8)	701	803	936	1105	1333	1645	2128	2930	4455	8426
	(0.8/0.2)	705	815	957	1149	1412	1806	2433	3603	6298	17671

References

1. Tchanche, B.F.; Lambrinos, G.; Frangoudakis, A.; Papadakis, G. Low-grade heat conversion into power using organic Rankine cycles—A review of various applications. *Renew. Sustain. Energy Rev.* **2011**, *15*, 3963–3979. [CrossRef]
2. Quoilin, S.; van den Broek, M.; Declaye, S.; Dewallef, P.; Lemort, V. Techno-economic survey of Organic Rankine Cycle (ORC) systems. *Renew. Sustain. Energy Rev.* **2013**, *22*, 168–186. [CrossRef]
3. Colonna, P.; Casati, E.; Trapp, C.; Mathijssen, T.; Larjola, J.; Turunen-Saaresti, T.; Uusitalo, A. Organic Rankine Cycle Power Systems: From the Concept to Current Technology, Applications, and an Outlook to the Future. *J. Eng. Gas Turbines Power* **2015**, *137*, 100801. [CrossRef]
4. Andreasen, J.; Meroni, A.; Haglind, F. A Comparison of Organic and Steam Rankine Cycle Power Systems for Waste Heat Recovery on Large Ships. *Energies* **2017**, *10*, 547. [CrossRef]
5. Demuth, O.; Kochan, R.J. *Analyses of Mixed Hydrocarbon Binary Thermodynamic Cycles For Moderate Temperature Geothermal Resources Using Regeneration Techniques*; Technical Report; US Department of Energy: Idaho Falls, ID, USA, December 1981.
6. Angelino, G.; Colonna, P. Multicomponent working fluids for organic Rankine cycles (ORCs). *Energy* **1998**, *23*, 449–463. [CrossRef]
7. Modi, A.; Haglind, F. A review of recent research on the use of zeotropic mixtures in power generation systems. *Energy Convers. Manag.* **2017**, *138*, 603–626. [CrossRef]
8. Radermacher, R. Thermodynamic and heat transfer implications of working fluid mixtures in Rankine cycles. *Int. J. Heat Fluid Flow* **1989**, *10*, 90–102. [CrossRef]
9. Heberle, F.; Preißinger, M.; Brüggemann, D. Zeotropic mixtures as working fluids in Organic Rankine Cycles for low-enthalpy geothermal resources. *Renew. Energy* **2012**, *37*, 364–370. [CrossRef]

10. Andreasen, J.G.; Larsen, U.; Knudsen, T.; Pierobon, L.; Haglind, F. Selection and optimization of pure and mixed working fluids for low grade heat utilization using organic Rankine cycles. *Energy* **2014**, *73*, 204–213. [CrossRef]

11. Lecompte, S.; Ameel, B.; Ziviani, D.; van den Broek, M.; De Paepe, M. Exergy analysis of zeotropic mixtures as working fluids in Organic Rankine Cycles. *Energy Convers. Manag.* **2014**, *85*, 727–739. [CrossRef]

12. Braimakis, K.; Preißinger, M.; Brüggemann, D.; Karellas, S.; Panopoulos, K. Low grade waste heat recovery with subcritical and supercritical Organic Rankine Cycle based on natural refrigerants and their binary mixtures. *Energy* **2015**, *88*, 80–92. [CrossRef]

13. Trapp, C.; Colonna, P. Efficiency Improvement in Precombustion CO_2 Removal Units with a Waste-Heat Recovery ORC Power Plant. *J. Eng. Gas Turbines Power* **2013**, *135*, 042311. [CrossRef]

14. Weith, T.; Heberle, F.; Preißinger, M.; Brüggemann, D. Performance of Siloxane Mixtures in a High-Temperature Organic Rankine Cycle Considering the Heat Transfer Characteristics during Evaporation. *Energies* **2014**, *7*, 5548–5565. [CrossRef]

15. Baik, Y.J.; Kim, M.; Chang, K.C.; Lee, Y.S.; Yoon, H.K. A comparative study of power optimization in low-temperature geothermal heat source driven R125 transcritical cycle and HFC organic Rankine cycles. *Renew. Energy* **2013**, *54*, 78–84. [CrossRef]

16. Baik, Y.J.; Kim, M.; Chang, K.C.; Lee, Y.S.; Yoon, H.K. Power enhancement potential of a mixture transcritical cycle for a low-temperature geothermal power generation. *Energy* **2012**, *47*, 70–76. [CrossRef]

17. Bombarda, P.; Invernizzi, C.M.; Pietra, C. Heat recovery from Diesel engines: A thermodynamic comparison between Kalina and ORC cycles. *Appl. Therm. Eng.* **2010**, *30*, 212–219. [CrossRef]

18. Le, V.L.; Kheiri, A.; Feidt, M.; Pelloux-Prayer, S. Thermodynamic and economic optimizations of a waste heat to power plant driven by a subcritical ORC (Organic Rankine Cycle) using pure or zeotropic working fluid. *Energy* **2014**, *78*, 622–638. [CrossRef]

19. Feng, Y.; Hung, T.; Zhang, Y.; Li, B.; Yang, J.; Shi, Y. Performance comparison of low-grade ORCs (organic Rankine cycles) using R245fa, pentane and their mixtures based on the thermoeconomic multi-objective optimization and decision makings. *Energy* **2015**, *93*, 2018–2029. [CrossRef]

20. Feng, Y.; Hung, T.; Greg, K.; Zhang, Y.; Li, B.; Yang, J. Thermoeconomic comparison between pure and mixture working fluids of organic Rankine cycles (ORCs) for low temperature waste heat recovery. *Energy Convers. Manag.* **2015**, *106*, 859–872. [CrossRef]

21. Heberle, F.; Brüggemann, D. Thermo-Economic Evaluation of Organic Rankine Cycles for Geothermal Power Generation Using Zeotropic Mixtures. *Energies* **2015**, *8*, 2097–2124. [CrossRef]

22. Oyewunmi, O.; Markides, C. Thermo-Economic and Heat Transfer Optimization of Working-Fluid Mixtures in a Low-Temperature Organic Rankine Cycle System. *Energies* **2016**, *9*, 448. [CrossRef]

23. Heberle, F.; Brüggemann, D. Thermo-Economic Analysis of Zeotropic Mixtures and Pure Working Fluids in Organic Rankine Cycles for Waste Heat Recovery. *Energies* **2016**, *9*, 226. [CrossRef]

24. Imran, M.; Usman, M.; Lee, D.H.; Park, B.S. Thermoeconomic analysis of organic Rankine cycle using zeotropic mixtures. In Proceedings of the 3rd International Seminar on ORC Power Systems, Brussels, Belgium, 12–14 October 2015; p. 161.

25. Andreasen, J.; Kærn, M.; Pierobon, L.; Larsen, U.; Haglind, F. Multi-Objective Optimization of Organic Rankine Cycle Power Plants Using Pure and Mixed Working Fluids. *Energies* **2016**, *9*, 322. [CrossRef]

26. Mclinden, M.O.; Radermacher, R. Methods for comparing the performance of pure and mixed refrigerants in the vapour compression cycle. *Int. J. Refrig.* **1987**, *10*, 318–325. [CrossRef]

27. Högberg, M.; Vamling, L.; Berntsson, T. Calculation methods for comparing the performance of pure and mixed working fluids in heat pump applications. *Int. J. Refrig.* **1993**, *16*, 403–413. [CrossRef]

28. Mathworks. *Matlab 2018b Documentation*; Technical Report; Mathworks: Natick, MA, USA, 2018.

29. Lemmon, E.W.; Huber, M.; McLinden, M. *NIST Standard Reference Database 23: Reference Fluid Thermodynamic and Transport Properties-REFPROP*, version 9.0; National Institute of Standards and Technology, Standard Reference Data Program: Gaithersburg, MD, USA, 2010.

30. Bell, I.H.; Wronski, J.; Quoilin, S.; Lemort, V. Pure and Pseudo-pure Fluid Thermophysical Property Evaluation and the Open-Source Thermophysical Property Library CoolProp. *Ind. Eng. Chem. Res.* **2014**, *53*, 2498–2508. [CrossRef]

31. Shah, R.K.; Sekulić, D.P. *Fundamentals of Heat Exchanger Design*; John Wiley & Sons, Inc.: Hoboken, NJ, USA, 2003. [CrossRef]

32. Coulson, J.; Richardson, J.; Backhurst, J. *Coulson and Richardson's Chemical Engineering*; Butterworth-Heinemann: Oxford, UK, 1999.

33. Walraven, D.; Laenen, B.; D'haeseleer, W. Optimum configuration of shell-and-tube heat exchangers for the use in low-temperature organic Rankine cycles. *Energy Convers. Manag.* **2014**, *83*, 177–187. [CrossRef]

34. Kærn, M.R.; Modi, A.; Jensen, J.K.; Haglind, F. An assessment of transport property estimation methods for ammonia-water mixtures and their influence on heat exchanger size. *Int. J. Thermophys.* **2015**, *36*, 1468–1497. [CrossRef]

35. Bell, K.J. *Final Report of the Cooperative Research Program on Shell-and-Tube Heat Exchangers*; Technical Report; University of Delaware: Newark, DE, USA, 1963.

36. Bell, K.J. Delaware method for shell design. In *Heat Transfer Equipment Design*; Shah, R.K., Subbarao, E.C., Mashelkar, R.A., Eds.; Hemisphere Publishing: Washington, DC, USA, 1988.

37. Taborek, J. Shell-and-tube heat exchangers: Single phase flow. In *Handbook of Heat Exchanger Design*; Hewitt, G.F., Ed.; Begell House: New York, NY, USA, 1998; Chapter 3.

38. Shah, M.M. An Improved and Extended General Correlation for Heat Transfer During Condensation in Plain Tubes. *HVAC&R Res.* **2009**, *15*, 889–913. [CrossRef]

39. Gnielinski, V. New equation for heat and mass transfer in turbulent pipe and channel flow. *Int. Chem. Eng.* **1976**, *16*, 359–368.

40. Petukhov, B. Heat Transfer and Friction in Turbulent Pipe Flow with Variable Physical Properties. *Adv. Heat Transf.* **1970**, *6*, 503–564. [CrossRef]

41. Bell, K.; Ghaly, M. An approximate generalized design method for multicomponent/partial condenser. *AIChE Symp. Ser.* **1973**, *69*, 72–79.

42. Blasius, H. Das Ähnlichkeitsgesetz bei Reibungsvorgängen in Flüssigkeiten. In *Mitteilungen über Forschungsarbeiten auf dem Gebiete des Ingenieurwesens 131*; Springer: Berlin, Germany, 1913.

43. Müller-Steinhagen, H.; Heck, K. A simple friction pressure drop correlation for two-phase flow in pipes. *Chem. Eng. Process. Process Intensif.* **1986**, *20*, 297–308. [CrossRef]

44. Martin, H. The generalized Lévêque equation and its practical use for the prediction of heat and mass transfer rates from pressure drop. *Chem. Eng. Sci.* **2002**, *57*, 3217–3223. [CrossRef]

45. Astolfi, M.; Romano, M.C.; Bombarda, P.; Macchi, E. Binary ORC (Organic Rankine Cycles) power plants for the exploitation of medium-low temperature geothermal sources—Part B: Techno-economic optimization. *Energy* **2014**, *66*, 435–446. [CrossRef]

46. Martelli, E.; Capra, F.; Consonni, S. Numerical optimization of Combined Heat and Power Organic Rankine Cycles—Part A: Design optimization. *Energy* **2015**, *90*, 310–328. [CrossRef]

47. Meroni, A.; Andreasen, J.G.; Persico, G.; Haglind, F. Optimization of organic Rankine cycle power systems considering multistage axial turbine design. *Appl. Energy* **2018**, *209*, 339–354. [CrossRef]

48. Garg, P.; Kumar, P.; Srinivasan, K.; Dutta, P. Evaluation of isopentane, R-245fa and their mixtures as working fluids for organic Rankine cycles. *Appl. Therm. Eng.* **2013**, *51*, 292–300. [CrossRef]

49. Garg, P.; Kumar, P.; Srinivasan, K.; Dutta, P. Evaluation of carbon dioxide blends with isopentane and propane as working fluids for organic Rankine cycles. *Appl. Therm. Eng.* **2013**, *52*, 439–448. [CrossRef]

50. Liu, W.; Meinel, D.; Gleinser, M.; Wieland, C.; Spliethoff, H. Optimal Heat Source Temperature for thermodynamic optimization of sub-critical Organic Rankine Cycles. *Energy* **2015**, *88*, 897–906. [CrossRef]

51. Andreasen, J.G.; Larsen, U.; Haglind, F. Design of organic Rankine cycles using a non-conventional optimization approach. In Proceedings of the ECOS 2015—the 28th International Conference on Efficiency, Cost, Optimization, Simulation and Environmental Impact of Energy Systems, Pau, France, 30 June–3 July 2015.

52. Gani, R.; Nielsen, B.; Fredenslund, A. A group contribution approach to computer-aided molecular design. *AIChE J.* **1991**, *37*, 1318–1332. [CrossRef]

53. Doucet, J.P.; Weber, J. *Computer-Aided Molecular Design: Theory and Applications*; Academic Press: Cambridge, MA, USA, 1996; p. 487.

54. Papadopoulos, A.I.; Stijepovic, M.; Linke, P. On the systematic design and selection of optimal working fluids for Organic Rankine Cycles. *Appl. Therm. Eng.* **2010**, *30*, 760–769. [CrossRef]

55. Lampe, M.; Stavrou, M.; Bücker, H.M.; Gross, J.; Bardow, A. Simultaneous Optimization of Working Fluid and Process for Organic Rankine Cycles Using PC-SAFT. *Ind. Eng. Chem. Res.* **2014**, *53*, 8821–8830. [CrossRef]

56. Lampe, M.; Stavrou, M.; Schilling, J.; Sauer, E.; Gross, J.; Bardow, A. Computer-aided molecular design in the continuous-molecular targeting framework using group-contribution PC-SAFT. *Comput. Chem. Eng.* **2015**, *81*, 278–287. [CrossRef]
57. Cignitti, S.; Andreasen, J.; Haglind, F.; Woodley, J.; Abildskov, J. Integrated working fluid-thermodynamic cycle design of organic Rankine cycle power systems for waste heat recovery. *Appl. Energy* **2017**, *203*. [CrossRef]

Article

Off-Design Performances of an Organic Rankine Cycle for Waste Heat Recovery from Gas Turbines

Carlo Carcasci *, Lapo Cheli, Pietro Lubello and Lorenzo Winchler

DIEF—Department of Industrial Engineering, University of Florence Via Santa Marta, 3, 50139 Florence, Italy; lapo.cheli@unifi.it (L.C.); pietro.lubello@unifi.it (P.L.); lorenzo.winchler@htc.de.unifi.it (L.W.)
* Correspondence: carlo.carcasci@unifi.it

Received: 24 January 2020; Accepted: 26 February 2020; Published: 2 March 2020

Abstract: This paper presents an off-design analysis of a gas turbine Organic Rankine Cycle (ORC) combined cycle. Combustion turbine performances are significantly affected by fluctuations in ambient conditions, leading to relevant variations in the exhaust gases' mass flow rate and temperature. The effects of the variation of ambient air temperature have been considered in the simulation of the topper cycle and of the condenser in the bottomer one. Analyses have been performed for different working fluids (toluene, benzene and cyclopentane) and control systems have been introduced on critical parameters, such as oil temperature and air mass flow rate at the condenser fan. Results have highlighted similar power outputs for cycles based on benzene and toluene, while differences as high as 34% have been found for cyclopentane. The power output trend with ambient temperature has been found to be influenced by slope discontinuities in gas turbine exhaust mass flow rate and temperature and by the upper limit imposed on the air mass flow rate at the condenser as well, suggesting the importance of a correct sizing of the component in the design phase. Overall, benzene-based cycle power output has been found to vary between 4518 kW and 3346 kW in the ambient air temperature range considered.

Keywords: ORC integration technologies; advanced thermodynamic cycles; decentralised energy systems; benzene; toluene; cyclopentane

---

1. Introduction

In recent years, the accelerated consumption of fossil fuels caused many serious environmental problems, such as global warming, depletion of the ozone layer and atmospheric pollution. A statistic survey from Hung [1] demonstrated that more than 50% of waste heat produced by industries is at low temperatures and therefore is difficult to recover with conventional systems based on Rankine steam cycle.

Organic Rankine Cycle (ORC) power plants have proved to be an attractive solution for the conversion of low/medium grade heat into electricity. Furthermore, this kind of cycle allows the usage of clean energy sources, such as solar radiation [2], geothermal water/steam [3,4], biomass [5] and waste combustion. ORC works on organic fluids, which usually have higher molecular weight and lower evaporation points with respect to water: this permits the increase of the efficiency when working below 400–500 °C. Dry fluids are preferred because of the positive slope of their saturation vapour curve that prevents liquid formation during expansion [6]. This choice is also based on environmental impact: several studies on the fluid's polluting properties can be found in literature, as in the work of Chen [7]. ORC power plants tend to have numerous advantages, such as reliability, quiet operation, long life and compact size [8].

Many studies can be found in the literature, in particular on thermodynamic analysis with several fluids and design optimisation: it has been observed that to avoid the formation of wet fraction expansion when the cycle is not overheated, the evaporation pressure must not exceed the pressure characterised by the maximum saturation specific entropy [9]. The most significant parameters influencing cycle performances are pinch point temperature difference, fluid flow and inlet evaporator diathermic oil temperature [10–14]. It should be noted that the expander, condenser and evaporator generate the greater part of the exergetic losses, which tend to increment with increasing expander inlet temperature and gas temperature [15,16].

Several papers have focused on combined cycles employing gas turbine as topper cycle and alternative solutions to the traditional steam-water Rankine cycle for the bottoming one. Chacartegui [17] proposed ORC as an alternative bottoming cycle for a combined cycle power plant, while Del Turco [18] proposed the so-called *ORegen cycle*. Carcasci [19] studied the ORC bottoming cycle at design conditions for different working fluids (benzene, toluene and cyclopentane), showing the best efficiency settings by varying fluid evaporation pressure and oil temperature, with a GE10 gas turbine [20]. More recent papers have studied the behaviour of Organic Rankine Cycle in off-design conditions, e.g., Quoilin [21] have studied a regulation dynamic model for transitional state and off-design. They proposed three different strategies for expander and pump control systems and have demonstrated that the most effective one is the evaporation temperature optimisation according to the working conditions. Song [22] analysed the behaviour of an ORC system using a parametric analysis and a one-dimensional analysis method for the aerodynamic model of the turbine, working both under design and off-design conditions; the inlet temperatures of the heat source and the cooling water were found to have a significant influence on the system performance. Cao [23] presents a comparative analysis on the off-design performance of a gas turbine and organic Rankine cycle (GT-ORC) combined cycle under different operation approaches; the GT-ORC cycle efficiency was maximised with a sliding pressure operation method in order to reduce the influence of ambient temperature on the overall efficiency.

Another dynamic control model was demonstrated by Wei [24], who obtained an accuracy of 4% without presenting numerical inaccuracies due to the oscillation phenomena. Manente [4] analyzed an ORC for geothermal heat recovery. The key elements of the dynamic model were two tanks, one at high and one at low pressure, which made the two main flows independent respectively to the expander and to the pump. By varying the ambient and geothermal fluid temperature, they have optimised the performance by changing the pump's revolutions per minute, turbine's capacity factor and condenser's refrigerant mass flow rate. The authors have identified a limit in the optimisation of the cycle: the geothermal fluid reinjection temperature should not be less than 70 °C. Sauret [25] performed a three-dimensional off-design numerical analysis of an ORC. First, they identified R143a as the best high density fluid to optimise the turbine's outlet power, then they analysed the radial-inflow turbine operation in several part-load conditions. Fu [26] investigated the cycle performance by varying the heat source temperature: an increment of this parameter generates an increase in net power and heat transfer capacity. Ibarra [27] computed the Organic Rankine Cycle performance at several imposed power outputs, using R245fa and SES36 fluids; by varying expansion speed and working pressure they obtained the off-design maximum efficiency.

Bamgbopa [28] employed an unsteady model to analyse the efficiency of a solar ORC while comparing different working fluid mass flow rates, heat source temperatures and heat source mass flow rates. De Escalona [29], in order to identify the best part-load control strategy of a gas turbine combined with ORC, has maintained the efficiency of the bottomer cycle at the design value by imposing a constant evaporating pressure. The topper cycle efficiency decreases with the load. When this happens the heat supplied to the ORC increases, resulting in a higher power output and better overall performances. Lecompte [30] developed a thermo-economic optimisation method minimising the investment specific cost at partial loads Specific Investment Cost Part Load (SICPL); SICPL is a polynomial expression depending on ambient temperature and supplied thermal power. The authors proved that investment specific costs increase from design conditions to variable thermodynamic ones. Calise [31] proposed another thermo-economic optimisation model, based on investment costs minimisation. In off-design conditions, they noted an increase in the cycle's power output in the presence of a fluid mass flow rate decrease. Finally, Quoilin [32] demonstrated that better performances can be obtained for evaporation temperatures significantly lower than the heat source one, while for higher values (ensuring higher pressures and densities) the economic optimum is obtained due to the smaller size of the heat exchanger.

The novelty of the presented work is the study of the off-design performances of an ORC with the inclusion of its control system; moreover, the ORC is fed by waste heat recovered from a gas turbine exhaust, employing a similar cycle to the one described by Carcasci [33]. The analysis has been conducted for three different working fluids (benzene, toluene and cyclopentane) and has been performed varying the ambient air temperature while monitoring the cycle's characteristic parameters. Related trends are identified and control limits imposed to avoid exceeding in working conditions the imposed boundaries for critical parameters.

2. Materials and Methods

2.1. Working Fluid

The considered working fluids are benzene, toluene and cyclopentane. When a working fluid for ORC application is chosen, several criteria need to be considered: environmental sustainability, safety, thermal stability and critical temperature. The employed fluids meet most of the criteria listed, indeed ORC cycles based on such fluids are widely analysed in the literature [17,18]. The three fluids are dry and belong to the hydrocarbons family: the first two are aromatic and the third is a cycloalkane. Typically, dry fluids are preferred, since superheating is not necessary in order to avoid condensation during vapor expansion. To simulate the fluids behaviour, a NIST software was employed (NIST4 in the specific application).

Critical temperatures reported in Table 1 are high enough to guarantee a wide saturation curve and great chemical stability at the analysed temperatures. Table 1 shows also high molar masses (and densities), which allow a high specific heat absorption for each fluid, thus reducing the required mass flow rate. This enables the reduction of power consumption, pump size and the required surface for the heat exchangers, therefore decreasing the overall plant size and its cost. Moreover, smaller mass flow rates imply reduced environmental impact.

Table 1. Thermodynamic properties of selected working fluids.

Working Fluid	T_{cr} (°C)	P_{cr} (bar)	M_m (g/mol)	$P_{s,max}$ (bar)
Toluene	318.6	41.1	92.14	36.0
Benzene	288.8	48.9	78.10	37.5
Cyclopentane	238.6	44.3	70.13	34.7

The often used pollution indices are the Global Warming Potential (GWP) and Ozone Depletion Potential (ODP). Hydrocarbons have low GWP (between 3 and 24 units in the case of gas methane) and zero ODP, since they do not contain chlorine or bromine atoms. Another positive aspect is the low electrical conductivity, which allows the fluids to be used as lubricant for the turbine's bearings and as coolant for the generator. On the other hand, some of these fluids have negative effects on human health: at ambient conditions, benzene is volatile, colourless, highly flammable, irritant and carcinogenic on long-term exposure; toluene is irritant as well, but less volatile and toxic than benzene; cyclopentane is the most flammable of the three fluids, with a visibly lower auto ignition temperature of 380 °C but a reduced adverse effect on human health, as demonstrated by the higher Threshold Limit Value (TLV). Saturation temperature value, as determined by *NIST4* library, has been plotted as a function of pressure for each organic fluid and is reported in Figure 1: toluene shows the higher saturation temperature regardless of the pressure considered, while cyclopentane shows the lowest.

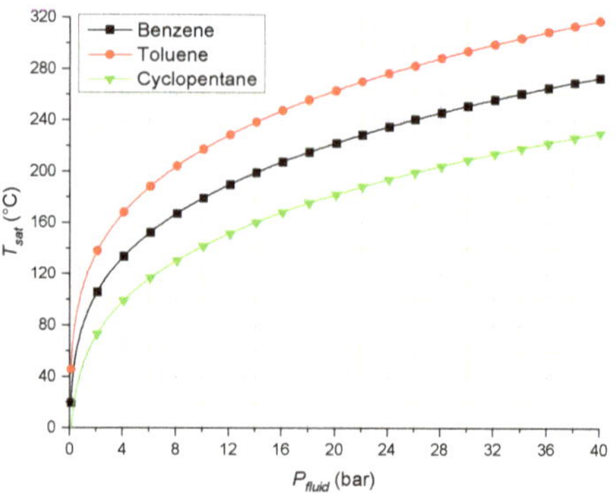

Figure 1. Saturation temperature vs. pressure for each organic fluid.

2.2. Power Plant Layout

The examined plant is based on the traditional subcritical Rankine cycle, combined with a gas turbine topping cycle. Figure 2 shows the power plant layout of the ORC bottoming cycle [19].

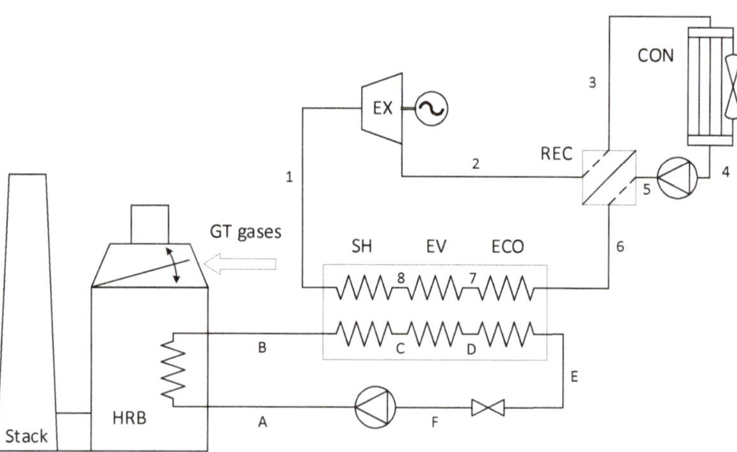

Figure 2. Organic Rankine Cycle (ORC) power plant layout. Points 1–8: organic fluid cycle, points A–F: diathermic oil cycle

The gas turbine is the GE10-1 from General Electric–Nuovo Pignone; it is a heavy-duty single-shaft gas turbine used for power generation applications [20]. Table 2 lists the main specifications of the considered gas turbine. For safety reasons, the recovery cycle is designed for a heat transfer made by an intermediate diathermic oil circuit placed between the hot gas turbine exhaust flow and the organic fluid loop.

Table 2. GE10 single shaft gas turbine main specification [20].

W_{GT} (kW)	η_{GT} (%)	m_{EXH} (kg/s)	T_{EXH} (°C)
11,250	31.4	47.5	482.0

The cycle has only one pressure level and, in order to increase the system efficiency, a recuperative heat exchanger (REC) has been introduced. The hot exhaust gas mass flow (first cycle) heats the diathermic oil (second cycle, points A to F) in the heat recovery boiler (HRB), the hot oil passes then through the heat recovery steam generator (points B-E), composed by economizer (ECO), evaporator (EV) and superheater (SH). A third separated cycle is the organic fluid loop (points 1–8). After leaving the HRSG, the superheated fluid enters the turbo-expander (EX). The exhaust organic fluid exchanges then heats in the recuperator (REC), where the condensed fluid is heated. Finally, the organic fluid is cooled in a condenser (CON) and pressurised in the extraction pump. The condenser is of an air-cooled type, as the application in water-scarce areas has been considered.

2.3. Off-Design Analysis

When the plant works in off-design conditions, the temperature difference between the hot and cold fluid cannot be imposed, while the geometry (heat exchange surfaces) is the same obtained in the design phase. The surface is determined in the design condition by imposing the temperature difference, energy balances and mass flow rates. Through the logarithmic mean temperature difference the overall heat transfer coefficient UA value in design condition can be determined:

$$UA_{des} = \frac{Q}{\Delta T_{ml}} \tag{1}$$

In the Heat Recovery Boiler (HRB), temperature and mass flow rate of gas turbine's exhaust gas are imposed (m_{exh}, T_{exh}; see Table 2 and Figure 3) and maximum oil temperature is limited by the oil properties: $T_B = T_{oil,max} = 380\,°\text{C}$.

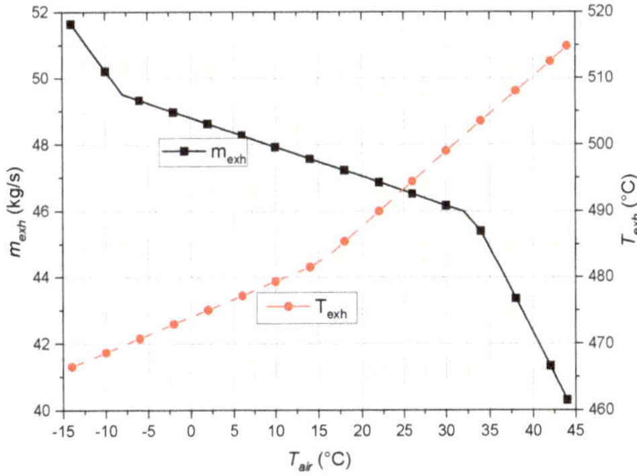

Figure 3. Mass flow rate and temperature exhaust from GE10 gas turbine, varying ambient temperature.

Contrary to design analysis, pinch point cannot be imposed, but the UA_{HRB} in gas-oil heat recovery boiler is imposed instead, and by using the energy balance, stack temperature of the hot gas stream can be determined. This value must be higher than stack temperature limit—$T_{st,lim} = 90\,°\text{C}$—to avoid the occurrence of corrosion phenomena. Moreover, the oil mass flow rate (m_{oil}) can be determined: the control system sets the maximum oil temperature that will be obtained by varying the oil valve position and pump settings, thus allowing the oil mass flow rate to vary consequently.

On the other side of the plant, considering that ambient conditions are known (even though different from the design case), employing the energy balance and the overall heat transfer coefficient of the condenser (UA_{con}), the outlet air temperature ($T_{air,out}$) can be determined. This can be done by dividing the condenser in two sections, one for the condensation and one for the vapour cooling, and by imposing the organic fluid saturation pressure ($P_4 = P_{con}$). Furthermore, the air mass flow rate of the fan condenser (m_{air}) is determined using the energy balance in the condenser. The power absorbed by the fan (W_{fan}) can be determined as well:

$$W_{fan} = m_{con,air} \frac{\Delta P_{con,air}}{\rho_{air}} \tag{2}$$

In practice, the control system varies the fan speed to obtain the imposed condenser pressure (maintained constant), even though the fan and its electric motor have performance limits. The reduced mass flow rate in the expander and the exhaust pressure (from the condenser balance) can be determined but they depend on the expander characteristic. There are two different ways to proceed:

- Working with a fixed expander nondimensional flow, so that the inlet pressure can vary (the inlet pressure is set by the expander characteristic and the expander is often chocked, resulting in a constant reduced mass flow rate, compared to the design condition [9])
- Working with a control system: the analysed cycle has only one level of evaporating pressure with an internal heat exchanger, the recuperator (REC), to increase the system efficiency (valve) at the inlet of the expander that allows for the setting of the pressure

In any case, it is possible to start imposing the maximum fluid pressure and, together with the UA in the evaporator (EV) and the energy balance, the organic fluid mass flow rate can be calculated. The exhaust condition of the organic fluid from the $HRSG$ is considered to be saturated steam at maximum pressure. If the superheater (SH) is present, another energy balance is necessary and its UA_{SH} to obtain the exhaust organic fluid temperature. If the expander is equipped with an inlet pressure control system, it adjusts the expander to the inlet flow conditions. If the inlet pressure control system is not present, the mass flow rate and pressure at the inlet of the expander must match its characteristic curve. Supposing that the expander is chocked, the nondimensional flow is constant (and has been determined in design condition). The inlet pressure cannot be imposed and has to be obtained from the following equation:

$$\frac{m_{fl,off} \cdot \sqrt{T_{1,off}}}{P_{1,off}} = \frac{m_{fl} \cdot \sqrt{T_1}}{P_1} = constant \tag{3}$$

Thus, an iterative calculation is necessary between the evaporator (superheater, if present) and the expander, indeed the organic fluid mass flow rate and the maximum pressure are determined by matching these components. The isentropic efficiency depends on the expander operating conditions [31]. Once computed, it allows the organic fluid exhaust conditions to be obtained. Applying the energy balance and heat transfer equations at the recuperator (REC), the organic fluid conditions at the inlet of the condenser (CON) and economizer (ECO) can be determined. A new value for the return oil temperature and for the organic fluid inlet condition at the evaporator are computed, hence closing the oil and organic fluid loops.

The global electrical power output can be determined:

$$W_{el} = m_{fl} \cdot (L_{ex} - L_{pump,fl} - L_{pump,oil}) - W_{fan} \tag{4}$$

A mass balance between the design and off-design cases has to be introduced in order to verify the continuity between the mass flow rates. If the balance is not verified, a tank has to be considered to compensate the mass flow rate fluctuations in the system. In the present study, the mass conservation is not considered due to the introduction of a tank. The fluid-dependent design conditions (Table 3) considered have been chosen for having the best performance for each fluid (as shown by previous studies like [9,19] at ISO conditions). Table 4 lists all the physical quantities that have been imposed in the design phase independently from the fluid considered. Once the thermodynamic computation of the cycle has been performed, parameters such as the logarithmic mean temperature difference (ΔT_{ml}), UA for each heat exchanger, dimensionless mass flow rate and expander pressure ratio can be determined.

Table 3. ORC power plant data in design conditions—parameters depending on working fluid.

Working Fluid	P_{des} (bar)	$m_{fl,des}$ (kg/s)	$m_{oil,des}$ (kg/s)	$W_{wl,des}$ (kW)
Toluene	36.0	30.52	30.04	4389.9
Benzene	37.5	33.15	29.56	4504.5
Cyclopentane	44.0	22.71	32.95	3652.7

Table 4. ORC power plant data in design conditions—parameters not depending on working fluid.

Parameter	Value	Parameter	Value
$T_{oil,max}$	380.0 °C	$\Delta T_{pp,CON}$	20.0 °C
$\Delta T_{pp,HRSG}$	8.0 °C	$\Delta T_{air,CON}$	20.0 °C
$\Delta T_{app,HRSG}$	30.0 °C	$\Delta T_{pp,REC}$	15.0 °C
$\Delta T_{sub,HRSG}$	15.0 °C	η_{HRB}	0.98
$\Delta T_{pp,HRB}$	10.0 °C	η_{EX}	0.85
$\Delta T_{sub,HRB}$	8.0 °C	η_{pump}	0.70

The authors developed an in-house code that is able to perform thermodynamic simulations of the proposed power plant. The code has been developed in ANSI Standard of the FORTRAN90 programming language and the elementary energy balances have been validated through the comparison with a commercial code.

2.4. Heat Exchanger HTC in Off-Design Conditions

In the previous paragraph, the UA in each section of heat recovery boiler and heat exchanger has been imposed. This value is calculated in design phase, but it changes in off-design conditions. In the present study, a simplified method to determine this value is proposed. The overall heat transfer coefficient is composed of three different parts, respectively addressing internal convection, external convection and conduction through the metal:

$$\frac{1}{U} = \frac{1}{U_{int}} + \frac{1}{k/s} + \frac{1}{U_{ext}} \tag{5}$$

Each thermal factor can be expressed as a fraction of the global thermal resistance

$$f_{int} = \frac{R_{int}}{R} = \frac{U}{U_{int}}$$
$$f_{cond} = \frac{R_{cond}}{R} = \frac{U}{k/s} \tag{6}$$
$$f_{ext} = \frac{R_{ext}}{R} = \frac{U}{U_{ext}}$$

to obtain

$$f_{int} + f_{cond} + f_{ext} = 1 \tag{7}$$

From prior knowledge of the mechanism of heat exchange characterising the considered heat exchanger, it is possible to determine the values of the addends f and, as a consequence, of the global heat transfer coefficient in design conditions. Typically, f_{cond} lies between 0.02 and 0.05, while f_{int} and f_{ext} tend to be between 0.30 and 0.60, depending if describing a component with higher external (e.g., superheater) or internal convection (e.g., economizer). In off-design analysis, the heat transfer coefficient is a function of the mass flow rate and fluid properties. Usually, a correlation for the heat transfer coefficient can be defined through Nusselt and Reynolds numbers [8]. Typical values for the exponent a are between 0.5 and 0.8.

$$Nu = C \cdot Re^a \rightarrow U \propto m^a \rightarrow \frac{U_{off}}{U_{des}} \propto \left(\frac{m_{off}}{m_{des}} \right)^a \tag{8}$$

Thus, with the off-design mass flow rate, the heat transfer coefficient of each contribution can be determined and, by considering the conduction thermal resistance of the metal (k/s) to be constant, the overall heat transfer coefficient can be calculated through Equation (5).

2.5. Critical Conditions and Valve Settings

Control devices are necessary to allow the power plant control in off-design conditions. Indeed, some critical conditions must be avoided:

- Stack temperature must not descend below a certain limit temperature to avoid corrosion phenomena: $T_{st} \geq T_{st,lim} = 90\,^{\circ}\text{C}$;
- The oil mass flow rate cannot exceed an upper limit (generally referred to the design value): $\alpha_{oil,low} \leq m_{oil}/m_{oil,des} \leq \alpha_{oil,up}$;
- The condenser air mass flow rate cannot exceed a lower and upper limit (generally referred to the design value and depending on fan size, increasing this value raises the fan cost [18]): $m_{con,air}/m_{con,air,des} \leq \alpha_{air,up}$;
- In the recuperator (REC), the organic fluid in the hot section must not reach saturation at outlet (otherwise, transportation of a two-phase fluid into a tube can be challenging);
- In the economizer (ECO), the fluid at the outlet must be subcooled in order to avoid any presence of vapour fraction into the economizer.

Hence, valves must be introduced throughout the power plant to enforce these bounds. A valve can be described by its flow coefficient:

$$CV = \frac{m}{\rho} \cdot \sqrt{\frac{\rho}{\Delta P}} = \frac{m}{\sqrt{\rho \cdot \Delta P}} \tag{9}$$

In design condition, this value can be determined for each valve. Considering the valve in design condition to be not completely open, an oversized CV for each valve can be defined ($CV_{max} = \alpha_v \cdot CV_{des}$). A typical operating range for each valve can thus be imposed:

$$CV_{low} = \alpha_{v,low} \cdot CV_{max} \tag{10}$$

$$CV_{up} = \alpha_{v,up} \cdot CV_{max} \tag{11}$$

In off-design conditions, the mass flow rate and pressure drop through each valve can be calculated and a valve flow coefficient determined, thus the valve position can be defined. The off-design flow coefficient must respect the operating range of the valve ($CV_{low} \leq CV_{off} \leq CV_{up}$), if these bounds are not respected a valve of different size (smaller or bigger) must be chosen.

3. Results

The analysis has been carried out varying ambient air temperature from $-15\,^{\circ}\text{C}$ to $+45\,^{\circ}\text{C}$. As a consequence, three thermodynamic conditions are changing at the same time: the condenser refrigerant inlet air temperature and mass flow rate and temperature of the exhaust gas from the gas turbine.

Increasing ambient air temperature, gas turbine's output power and electrical efficiency decrease, together with a reduction of the exhaust hot gas mass flow rate and an increment in their temperature (Figure 3). When ambient air temperature is lower than $-8\,^{\circ}$C and greater than $32\,^{\circ}$C, the exhaust mass flow rate decreases sharply. At $+15\,^{\circ}$C, the exhaust gas temperature trend slightly changes its slope.

The exhaust gas temperature value is, for all ambient temperature considered, greater than the oil limit temperature. Hence, the oil temperature at the outlet of the HRB is imposed to be equal to its limit temperature ($T_B = 380\,^{\circ}$C) in each off-design case, and the oil mass flow rate can be determined. The electrical power output shows a similar trend for each fluid considered (Figure 4), indicating the best performances for benzene and toluene independently from the ambient air temperature considered. The power output of the bottoming cycle decreases when air temperature increases, with a more significant power drop for higher temperatures. The curve is visibly steeper also for ambient temperatures lower than $-8\,^{\circ}$C. The power output is strongly affected by the hot gas mass flow rate (that shows a similar trend, as it can be seen in Figure 3), while the effects of the hot gas temperature are less evident since the maximum oil temperature is imposed to be equal to its limit.

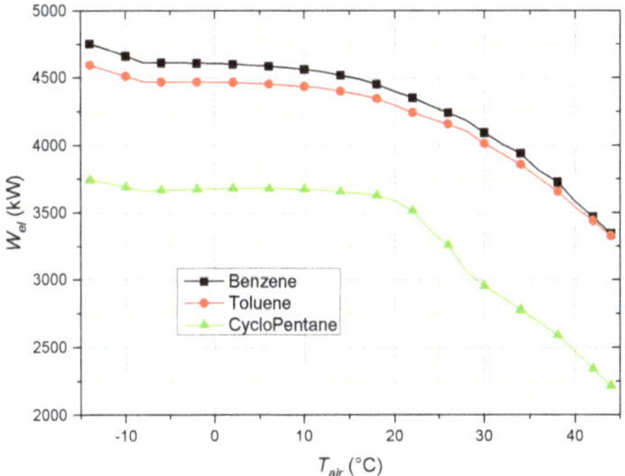

Figure 4. ORC electrical output power varying ambient temperature.

Considering Equation (4), each single contribution can be analysed. The power absorbed by the condenser fan (Figure 5) is not negligible; indeed, when ambient temperature increases, the control system, to maintain the pressure at the condenser constant, increases the air mass flow and, as a consequence, the pressure losses and power demand of the fan increase sharply (this can be clearly seen in Figure 5 for temperatures approaching $16\,^{\circ}$C). When the air mass flow rate limit in the fan occurs (Figure 6), the control system imposes the maximum mass flow rate, resulting in an almost constant power absorption for temperatures above $16\,^{\circ}$C (Figure 5), nonetheless a slight increase can still be appreciated, which is to be ascribed to the increase in the organic fluid pressure at the condenser. In the case of high ambient temperatures, an increase in the pressure at the condenser implies a higher saturation temperature for the organic fluid considered (Figure 1). The exhaust organic fluid temperature from the expander (T_2) tends to grow, as the expansion line is shorter. Consequently, the organic fluid temperature at the recuperator outlet (economizer inlet, T_6) is higher and the return oil temperature (T_E) increases as well.

A higher return oil temperature at HRB (T_A) reduces the heat exchanged with the hot gas stream inside the HRB, leading to an increase in the exhaust gas stack temperature (Figure 7) and maintaining it always above its lower limit. Figure 8 shows the diathermic oil mass flow rate varying with ambient air temperature. When air temperature is lower than $-8\,^{\circ}$C, the oil mass flow rate decreases, since the hot gas mass flow rate decreases as well (Figure 3). When ambient air temperature is between $-8\,^{\circ}$C and $16\,^{\circ}$C, the oil mass flow rate is almost constant, for the reduction of hot mass flow rate is compensated by the gain in temperature of the hot gas stream (Figure 3). When ambient air temperature rises above 16 °C, the return oil temperature increases and the oil mass flow rate increases again. Once the air temperature reaches 33 °C, the hot gas mass flow rate decreases sharply with respect to the increase in its temperature, leading to a decrease in the oil mass flow rate too.

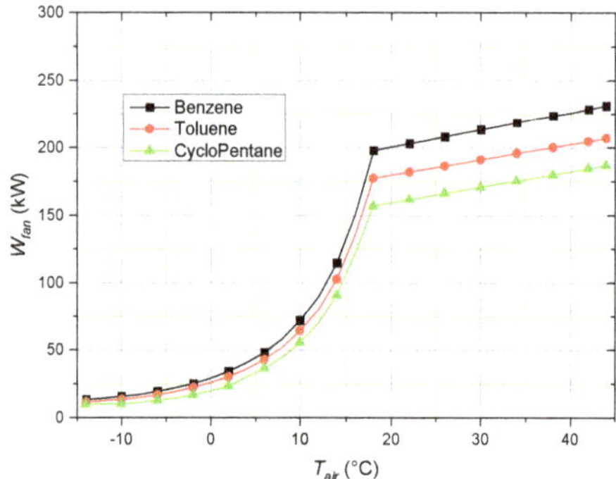

Figure 5. Power absorbed from condenser fan varying ambient temperature.

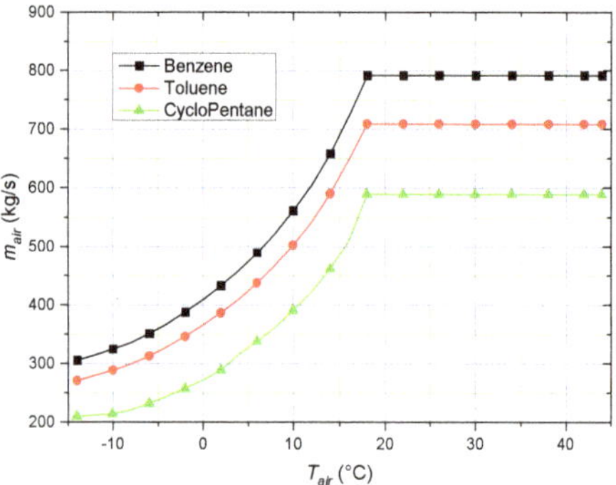

Figure 6. Condenser air mass flow rate varying ambient temperature.

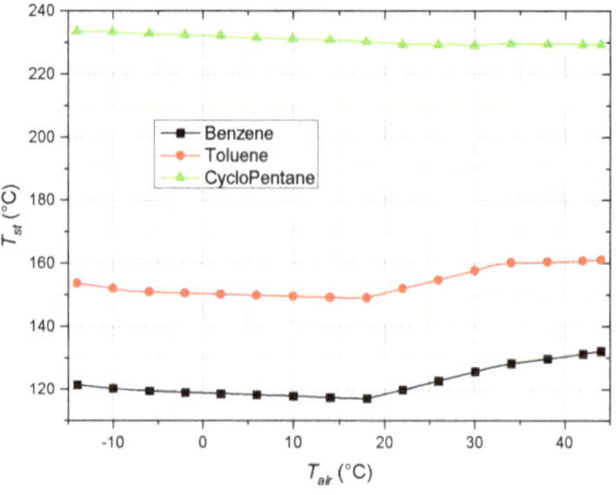

Figure 7. Stack temperature of exhaust gas varying ambient temperature.

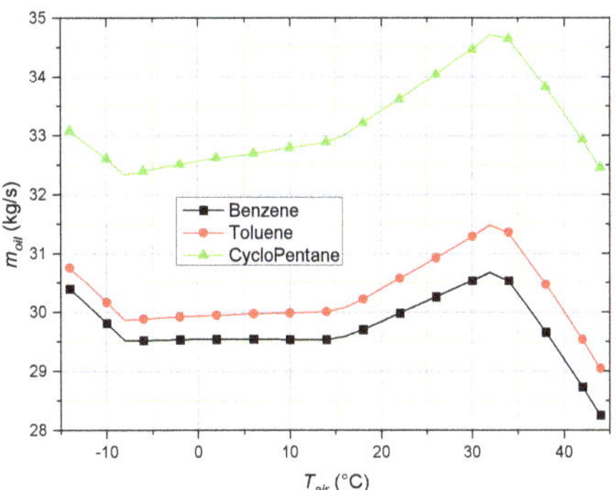

Figure 8. Diathermic oil mass flow rate varying ambient temperature.

Figure 9 shows the organic fluid mass flow rate as a function of ambient air temperature. The curves show a trend similar to those of the oil mass flow rates seen in Figure 8. Given the small magnitude of the subcooling inside the evaporator, from the energy balances it is possible to verify that the ratio between the fluid and oil mass flow rate is almost constant. The power absorbed from the organic fluid pump (Figure 10) presents quite the same trend of the organic fluid mass flow rate (Figure 9). The organic fluid mass flow rate variations are contained (lower than 8%), hence the expander inlet pressure oscillations are not relevant and show a trend similar to the fluid mass flow rate. Figure 11 presents the specific work per unit mass of the expander. It is almost constant for ambient air temperatures below 18 °C; however, when the condenser's fan reaches its mass flow rate limit and the pressure at the condenser still increases, the specific work will start decreasing.

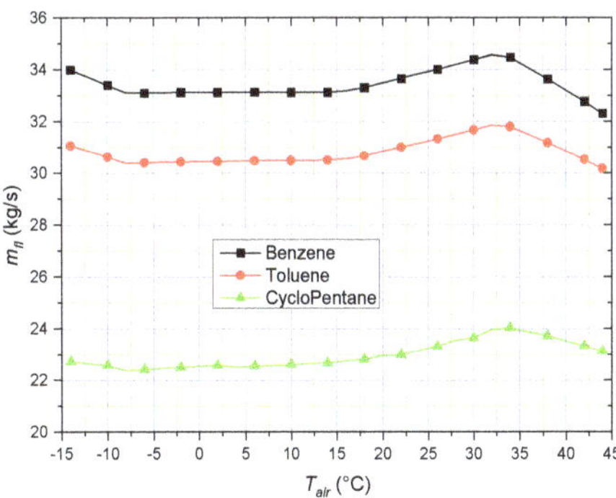

Figure 9. Organic fluid mass flow rate varying ambient temperature.

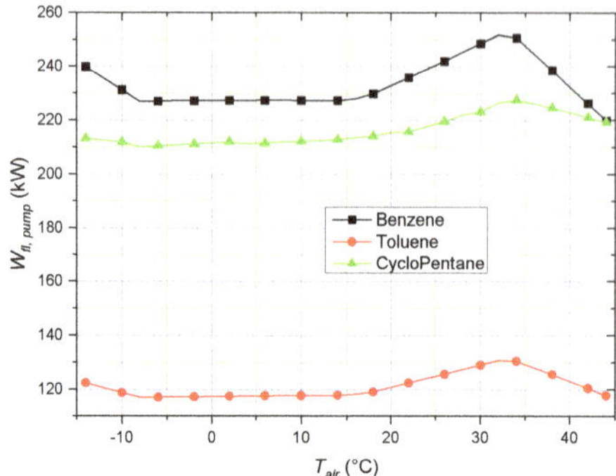

Figure 10. Power absorbed by organic fluid pump varying ambient temperature.

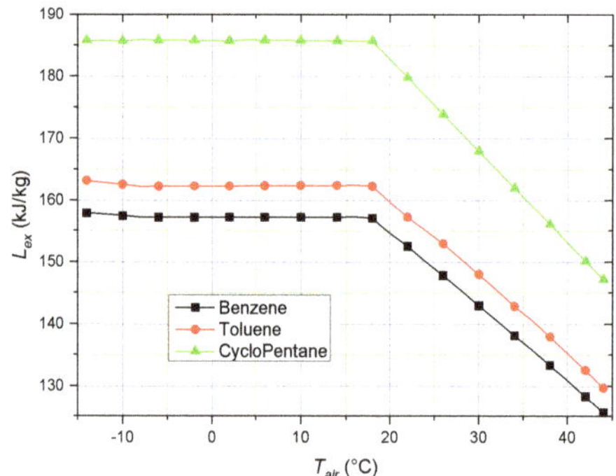

Figure 11. Expander specific work varying ambient temperature.

4. Conclusions

An Organic Rankine Cycle used as a bottoming cycle of a gas turbine has been investigated. An off-design performance analysis varying ambient air temperature (from $-15\,^{\circ}$C to $45\,^{\circ}$C) with three different organic fluids has been performed; all three fluids have shown similar behaviours, however, benzene and toluene have proved to be more performing than cyclopentane. The benzene-based cycle power output varies between 4518 kW and 3346 kW in the ambient air temperature range considered.

The electric power output of the ORC decreases when ambient air temperature increases. This trend is mainly due to two distinct effects: the gas turbine exhaust mass flow rate decrease and the maximum air mass flow rate of condenser fan. The former effect could be partially counterbalanced by the increase in the hot gas temperature, yet it results ineffective in doing so, since the maximum oil temperature is reached and imposed to be constant. The latter effect brings an increase in fluid pressure at the condenser: the expander specific work decreases and the heat recovered in the HRB decreases accordingly. These considerations highlight the importance of a correct sizing of the condenser in design conditions, since it will affect the off-design behaviour of the entire cycle.

Future developments within this research topic could extend the study of the analysed power plant under off-design conditions using mixed working fluids, such as a benzene-toluene binary mixture, which can ensure better temperature glide so as to identify the most thermodynamically beneficial working fluids.

Author Contributions: Conceptualization, C.C. and L.W.; Data curation, L.W.; Formal analysis, L.C. and P.L.; Project administration, C.C.; Software, L.W.; Supervision, C.C.; Validation, L.C. and P.L.; Writing–original draft, L.W.; Writing–review and editing, L.C. and P.L. All authors have read and agreed to the published version of the manuscript.

Funding: This research received no external funding.

Conflicts of Interest: The authors declare no conflict of interest.

Abbreviations

The following abbreviations and nomenclature are used in this manuscript:

Greek

α	Valve CV corrective coefficient
η	effectiveness
ρ	density

Abbreviations

CON	Condenser
ECO	Economizer
EV	Evaporator
EX	Turbo-Expander
GWP	Global Warming Potential
HRB	Heat Recovery Boiler
HRSG	Heat Recovery Steam Generator
HTC	Heat Exchange Coefficient
ODP	Ozone Depletion Potential
ORC	Organic Rankine Cycle
REC	Recuperator
SICPL	Specific Investment Cost Part Load
SH	Superheater
TLV	Threshold Limit Value

Subscripts

air	external air
app	approach point
con	condenser
des	design
ex	exit
exh	exhaust
ext	external
fan	condenser fan
fl	organic fluid
int	internal
max	maximum
ml	logarithmic mean
oil	diathermic oil loop
off	off-design
pp	pinch point
pump	circuit pump
st	stack
sub	sub cooling
up	upper limit
v	valve

Symbol

A	Heat exchanger area
a	Reynolds exponent
C	Coefficient
CV	Valve discharge coefficient
f	Friction factor
k	Thermal conductivity
m	Mass flow rate
Nu	Nusselt number
Q	Heat output
R	Thermal resistance
Re	Reynolds number
s	Heat exchanger thickness
U	Global heat exchange coefficient
W	Power output

References

1. Hung, T.C.; Shai, T.; Wang, S.K. A review of organic Rankine cycles (ORCs) for the recovery of low-grade waste heat. *Energy* **1997**, *22*, 661–667. [CrossRef]
2. Quoilin, S.; Orosz, M.; Hemond, H.; Lemort, V. Performance and design optimization of a low-cost solar organic Rankine cycle for remote power generation. *Sol. Energy* **2011**, *85*, 955–966. [CrossRef]
3. Schuster, A.; Karellas, S.; Kakaras, E.; Spliethoff, H. Energetic and economic investigation of Organic Rankine Cycle applications. *Appl. Therm. Eng.* **2009**, *29*, 1809–1817. [CrossRef]
4. Manente, G.; Toffolo, A.; Lazzaretto, A.; Paci, M. An Organic Rankine Cycle off-design model for the search of the optimal control strategy. *Energy* **2013**, *58*, 97–106, doi:10.1016/j.energy.2012.12.035. [CrossRef]
5. Łukowicz, H.; Kochaniewicz, A. Analysis of the use of waste heat obtained from coal-fired units in Organic Rankine Cycles and for brown coal drying. *Energy* **2012**, *45*, 203–212. [CrossRef]
6. Hung, T.; Wang, S.; Kuo, C.; Pei, B.; Tsai, K. A study of organic working fluids on system efficiency of an ORC using low-grade energy sources. *Energy* **2010**, *35*, 1403–1411. [CrossRef]
7. Chen, H.; Goswami, D.Y.; Stefanakos, E.K. A review of thermodynamic cycles and working fluids for the conversion of low-grade heat. *Renew. Sustain. Energy Rev.* **2010**, *14*, 3059–3067. [CrossRef]
8. Bhargava, R.K.; Bianchi, M.; Branchini, L.; De Pascale, A.; Orlandini, V. Organic rankine cycle system for effective energy recovery in offshore applications: A parametric investigation with different power rating gas turbines. In Proceedings of the ASME Turbo Expo 2015: Turbine Technical Conference and Exposition, Montreal, QC, Canada, 15–19 June 2015; p. V003T20A004.
9. Carcasci, C.; Ferraro, R. Thermodynamic Optimization and Off-Design Performance Analysis of a Toluene Based Rankine Cycle for Waste Heat Recovery from Medium-Sized Gas Turbines. In Proceedings of the ASME 2012 Gas Turbine India Conference, Mumbai, India, 1 December 2012; pp. 761–772.
10. Dai, Y.; Wang, J.; Gao, L. Parametric optimization and comparative study of organic Rankine cycle (ORC) for low grade waste heat recovery. *Energy Convers. Manag.* **2009**, *50*, 576–582. [CrossRef]
11. Roy, J.; Mishra, M.; Misra, A. Parametric optimization and performance analysis of a waste heat recovery system using Organic Rankine Cycle. *Energy* **2010**, *35*, 5049–5062. [CrossRef]
12. Wei, D.; Lu, X.; Lu, Z.; Gu, J. Performance analysis and optimization of organic Rankine cycle (ORC) for waste heat recovery. *Energy Convers. Manag.* **2007**, *48*, 1113–1119. [CrossRef]
13. Guo, C.; Du, X.; Yang, L.; Yang, Y. Performance analysis of organic Rankine cycle based on location of heat transfer pinch point in evaporator. *Appl. Therm. Eng.* **2014**, *62*, 176–186. [CrossRef]
14. Sun, J.; Li, W. Operation optimization of an organic Rankine cycle (ORC) heat recovery power plant. *Appl. Therm. Eng.* **2011**, *31*, 2032–2041. [CrossRef]
15. Chen, Q.; Xu, J.; Chen, H. A new design method for Organic Rankine Cycles with constraint of inlet and outlet heat carrier fluid temperatures coupling with the heat source. *Appl. Energy* **2012**, *98*, 562–573. [CrossRef]
16. Li, J.; Pei, G.; Li, Y.; Wang, D.; Ji, J. Energetic and exergetic investigation of an organic Rankine cycle at different heat source temperatures. *Energy* **2012**, *38*, 85–95. [CrossRef]

17. Chacartegui, R.; Sánchez, D.; Muñoz, J.; Sánchez, T. Alternative ORC bottoming cycles for combined cycle power plants. *Appl. Energy* **2009**, *86*, 2162–2170. [CrossRef]
18. Del Turco, P.; Asti, A.; Del Greco, A.S.; Bacci, A.; Landi, G.; Seghi, G. The ORegen™ waste heat recovery cycle: Reducing the CO2 footprint by means of overall cycle efficiency improvement. In Proceedings of the ASME 2011 Turbo Expo: Turbine Technical Conference and Exposition, Vancouver, BC, Canada, 6–10 June 2011; pp. 547–556.
19. Carcasci, C.; Ferraro, R.; Miliotti, E. Thermodynamic analysis of an organic Rankine cycle for waste heat recovery from gas turbines. *Energy* **2014**, *65*, 91–100. [CrossRef]
20. Gas Turbine World. *2016-2017 Gas Turbine World Handbook*, 1st ed.; Pequot Publishing Inc.: Fairfield, CT, USA, 2017; p. 51
21. Quoilin, S.; Aumann, R.; Grill, A.; Schuster, A.; Lemort, V.; Spliethoff, H. Dynamic modeling and optimal control strategy of waste heat recovery Organic Rankine Cycles. *Appl. Energy* **2011**, *88*, 2183–2190. [CrossRef]
22. Song, J.; Gu, C.W.; Ren, X. Parametric design and off-design analysis of organic Rankine cycle (ORC) system. *Energy Convers. Manag.* **2016**, *112*, 157–165, doi:10.1016/j.enconman.2015.12.085. [CrossRef]
23. Cao, Y.; Dai, Y. Comparative analysis on off-design performance of a gas turbine and ORC combined cycle under different operation approaches. *Energy Convers. Manag.* **2017**, *135*, 84–100, doi:10.1016/j.enconman.2016.12.072. [CrossRef]
24. Wei, D.; Lu, X.; Lu, Z.; Gu, J. Dynamic modeling and simulation of an Organic Rankine Cycle (ORC) system for waste heat recovery. *Appl. Therm. Eng.* **2008**, *28*, 1216–1224. [CrossRef]
25. Sauret, E.; Gu, Y. Three-dimensional off-design numerical analysis of an organic Rankine cycle radial-inflow turbine. *Appl. Energy* **2014**, *135*, 202–211. [CrossRef]
26. Fu, B.R.; Hsu, S.W.; Lee, Y.R.; Hsieh, J.C.; Chang, C.M.; Liu, C.H. Effect of off-design heat source temperature on heat transfer characteristics and system performance of a 250-kW organic Rankine cycle system. *Appl. Therm. Eng.* **2014**, *70*, 7–12. [CrossRef]
27. Ibarra, M.; Rovira, A.; Alarcón-Padilla, D.C.; Blanco, J. Performance of a 5 kWe Organic Rankine Cycle at part-load operation. *Appl. Energy* **2014**, *120*, 147–158. [CrossRef]
28. Bamgbopa, M.O.; Uzgoren, E. Numerical analysis of an organic Rankine cycle under steady and variable heat input. *Appl. Energy* **2013**, *107*, 219–228. [CrossRef]
29. de Escalona, J.M.; Sánchez, D.; Chacartegui, R.; Sánchez, T. Part-load analysis of gas turbine & ORC combined cycles. *Appl. Therm. Eng.* **2012**, *36*, 63–72.
30. Lecompte, S.; Huisseune, H.; Van den Broek, M.; De Schampheleire, S.; De Paepe, M. Part load based thermo-economic optimization of the Organic Rankine Cycle (ORC) applied to a combined heat and power (CHP) system. *Appl. Energy* **2013**, *111*, 871–881. [CrossRef]
31. Calise, F.; Capuozzo, C.; Carotenuto, A.; Vanoli, L. Thermoeconomic analysis and off-design performance of an organic Rankine cycle powered by medium-temperature heat sources. *Sol. Energy* **2014**, *103*, 595–609. [CrossRef]
32. Quoilin, S.; Declaye, S.; Tchanche, B.F.; Lemort, V. Thermo-economic optimization of waste heat recovery Organic Rankine Cycles. *Appl. Therm. Eng.* **2011**, *31*, 2885–2893. [CrossRef]
33. Carcasci, C.; Winchler, L. Thermodynamic analysis of an Organic Rankine Cycle for waste heat recovery from an aeroderivative intercooled gas turbine. *Energy Procedia* **2016**, *101*, 862–869. [CrossRef]

Article

Energy and Exergy Analysis of Different Exhaust Waste Heat Recovery Systems for Natural Gas Engine Based on ORC

Guillermo Valencia [1,*]**, Armando Fontalvo** [2]**, Yulineth Cárdenas** [3]**, Jorge Duarte** [1] **and Cesar Isaza** [4]

[1] Programa de Ingeniería Mecánica, Universidad del Atlántico, Carrera 30 Número 8-49, Puerto Colombia, Barranquilla 080007, Colombia; jorgeduarte@mail.uniatlantico.edu.co
[2] Research School of Electrical, Mechanical and Material Engineering, The Australian National University, ACT 2600, Australia; armando.fontalvo@anu.edu.au
[3] Departamento de Energía, Universidad de la Costa, Barranquilla 080002, Colombia; ycardena6@cuc.edu.co
[4] Programa de Ingeniería Mecánica, Universidad Pontificia Bolivariana, Medellín 050004, Colombia; cesar.isaza@upb.edu.co
* Correspondence: guillermoevalencia@mail.uniatlantico.edu.co; Tel.: +575-324-94-31

Received: 18 May 2019; Accepted: 18 June 2019; Published: 20 June 2019

Abstract: Waste heat recovery (WHR) from exhaust gases in natural gas engines improves the overall conversion efficiency. The organic Rankine cycle (ORC) has emerged as a promising technology to convert medium and low-grade waste heat into mechanical power and electricity. This paper presents the energy and exergy analyses of three ORC–WHR configurations that use a coupling thermal oil circuit. A simple ORC (SORC), an ORC with a recuperator (RORC), and an ORC with double-pressure (DORC) configuration are considered; cyclohexane, toluene, and acetone are simulated as ORC working fluids. Energy and exergy thermodynamic balances are employed to evaluate each configuration performance, while the available exhaust thermal energy variation under different engine loads is determined through an experimentally validated mathematical model. In addition, the effect of evaporating pressure on the net power output, thermal efficiency increase, specific fuel consumption, overall energy conversion efficiency, and exergy destruction is also investigated. The comparative analysis of natural gas engine performance indicators integrated with ORC configurations present evidence that RORC with toluene improves the operational performance by achieving a net power output of 146.25 kW, an overall conversion efficiency of 11.58%, an ORC thermal efficiency of 28.4%, and a specific fuel consumption reduction of 7.67% at a 1482 rpm engine speed, a 120.2 L/min natural gas flow, 1.784 lambda, and 1758.77 kW of mechanical engine power.

Keywords: energy analysis; exergy analysis; organic Rankine cycle; waste heat recovery; natural gas engine

---

1. Introduction

The technological advances developed in organic Rankine cycles (ORC) applied to waste heat recovery (WHR) systems could become a promising feature for the engine manufacturing industry due to its capacity to reduce fuel consumption, increase net power output, and reduce greenhouse gas emissions [1].

ORC is considered as a feasible tool to increase overall conversion efficiency in industrial processes due to its capacity to recover energy from alternative sources, such as exhaust gases, cooling water, or lubricating oil, by using organic working fluids [2]. Furthermore, ORC configuration can be modified to maximize overall engine–ORC system performance by optimizing net power output, first law, and exergy efficiencies and minimizing exergy destruction [3]. Nevertheless, ORC–engine coupling must be carefully designed to avoid safety, performance, and revenue issues such as gas–fluid contact, as well as weight, complexity, and backpressure increase [4].

ORC–WHR research has addressed the integration between ORC and combustion engines. Plenty of studies have established that ORC improves the overall conversion efficiency by increasing net power production without penalizing fuel consumption. Patel and Doyle [5] presented a first attempt for WHR from diesel engines by using ORC. Their ORC system achieved an overall power increase of 13% in a Mack 676 diesel vehicle engine without increasing fuel consumption. Peris et al. [6] simulated six ORC configurations for WHR from cooling water in internal combustion engines (ICE) by using 10 non-flammable fluids. Their study showed that ICE electric efficiency could be increased by 4.9–5.3%, by achieving overall conversion efficiencies up to 7.15% at a relatively low-temperature cooling water (90 °C). Yu et al. [7] simulated a diesel engine–ORC integration for WHR from the engine exhaust gases and cooling system by using R245fa as the ORC working fluid. Their results showed that 75% of exhaust gases energy and 9.5% of cooling water energy could be recovered if ORC operating conditions are optimized and controlled to maintain the power output. However, these results are limited to an exergetic analysis of a single ORC configuration. Lu et al. [8] proposed an integrated diesel internal combustion engine and simple ORD (SORC) with solids adsorption technology for waste heat recovery from the cooling system and the engine exhaust gases. Their results showed a maximum recoverable power from the cooling process and exhaust systems of 67.9 kW and 82.7 kW, respectively.

Vaja and Gambarotta [9] evaluated the performance of SORC and ORC with a recuperator (RORC) configurations for WHR in a stationary 2.9-MW ICE at a single operation point. The results showed an increase of 12% in the overall efficiency of the system. Kalina [10] investigated the performance of a biomass power generation plant. The plant consisted of two natural gas ICE coupled to a biomass gasifier and an ORC. The ORC system was used as a WHR system to produce power from the engine's exhaust line and cooling system. Mingshan et al. [11] analyzed a combined heavy-duty diesel engine–SORC system for WHR. This system achieved a heat recovery efficiency between 10–15% when the heat exchanger operation is optimized. This publication also evaluated the engine operation under partial load conditions with a medium–high power condition. They concluded that variations in engine rotational speed must be determined to evaluate the true performance of the combined system.

Regarding ORC and stationary compressed natural gas (CNG) engines integration, two approaches have been studied: the use of multiple temperature loops for WHR from the engine intercooler, cooling system, and exhaust gases, and the use of single-temperature loops for WHR from engine exhaust. Within the first approach, Yao et al. [12] were the first ones to propose an ORC system for WHR from the engine intercooler, exhaust gases, and cooling system by using a low and a high-temperature loop. Their ORC system used R245fa and achieved 10.8% thermal efficiency and 26.9 kW of power, which increased overall power production by 33.7%, keeping the same fuel consumption. Yang et al. [13] optimized Yao's ORC configuration by using a genetic algorithm, but their reported performance was lower than Yao's results: 8.8–10.2% thermal efficiency and 23.6-kW net power output. Yao's configuration was studied again by Yang et al. [14] and analyzed the thermodynamic performance of the ORC with double-pressure configuration (DORC) operating with six working fluid groups integrated into a compressed natural gas engine, but performance indicators such as the waste heat recovery efficiency and specific fuel consumption were not considered. Wang et al. [15] also studied a dual-loop ORC system for WHR in stationary CNG engines but using R1233zd and R1234yf as working fluids and two recovery heat exchangers to increase ORC efficiency. The results showed that ORC efficiency was 10–14% when R1233zd was used in the high-temperature loop, and R1234yf was used in the low-temperature loop.

Within the second approach, Han et al. [16] carried out a dynamic simulation of an ORC with a double piston expander and R245ca as a working fluid. Their system took advantage of the engine exhaust gases to produce power in the ORC piston expander and compress the engine natural gas. However, they focused on the component efficiencies and compression benefits, but their study did not report overall efficiency values. Song et al. [17] proposed an RORC working with R416A as the

ORC fluid. Their results showed that this approach increased the overall conversion efficiency by 6%. Liu et al. [18] examined the effect of engine load on the performance of an ORC cycle combined with a stationary motor and R245fa as the working fluid. Their results showed ORC efficiencies between 16.3–25.9% when the engine load varies between 40–100%, respectively.

The selection of ORC working fluid is a key design step because it defines the operation limits and the power production potential. However, publications have shown that working fluid selection depends on the heat source and temperature limits. Chacartegui et al. [19] studied a low-temperature SORC configuration in medium and large-scale combined cycle plants for WHR from commercial gas turbines. Their study considered R-113, R-245, isobutene, toluene, cyclohexane, and isopentane as working fluids. The results showed that toluene and cyclohexane achieved the highest combined cycle efficiency—up to 60%–which is high global efficiency in this process. Qiu et al. [20] examined the experimental integration of a small-scale thermoelectric generation with a dual ORC. Drescher and Brüggemann [21] performed a screening of suitable organic working fluids for biomass-fired applications. Their selection involved critical temperature and pressure, dryness, turbine and pump efficiencies, and autoignition temperature. However, they obtained a group of suitable fluids rather than an optimum working fluid. Mago et al. [22] studied the effect of working fluid selection on the ORC performance at different heat source temperatures, supporting Drescher and Brüggemann's conclusions.

Kosmadakis et al. [23] tested more than 30 organic fluids and stated that R245fa is a suitable working fluid for ORC applications with ICE in terms of net power and thermal efficiency, but its use is restricted as international standards promotes low global warming potential (GWP) organic substances. Regarding solar thermal energy conversion, Tchance et al. [24] established that R134a is a suitable fluid due to its low toxicity and flammability, in addition to the high-pressure ratio and efficiency that can be achieved. Tian et al. [25] performed a energy analysis of a combined diesel ICE–SORC system to evaluate 20 different working fluids in terms of net output power and thermal efficiency. The study showed that R-141b showed the highest net power output (60 kJ/kg), but R-123 achieved the highest thermal efficiency (16.60%). Hung et al. [26] investigated an SORC integrated into a solar ventilation system. Their results showed an overall efficiency increase of 6.2%. Zare [27] evaluated the revenue of three ORC geothermal plants. They showed that the RORC configuration improves thermal efficiency, but requires additional components, increasing the total exergy destroyed, and affecting the overall exergy efficiency. This trend was confirmed by Fontalvo et al. [28], who evaluated three ORC configurations for low-temperature WHR and showed that SORC had the highest revenue because the additional equipment increases the exergy destruction of the system.

Exergy analysis is an important tool to identify key design aspects that may improve overall conversion efficiency and maximize resource utilization. ORC research has shown that exergy analysis can be used to provide guidelines in ORC design for a wide range of heat sources. In addition, the calculation of exergoeconomic costs aims to set an economic value for materials and energy flows, providing a reasonable base for the allocation of economic resources [29]. Kerme and Orfi [30] evaluated the effect of the turbine inlet temperature on the exergy efficiency of an ORC driven by solar collectors. As a result, they showed that an increase in turbine inlet temperature increases exergy efficiency and reduces exergy destruction. The conservation of natural resources, the limited availability of spaces to generate energy through some renewable sources, cost savings, policies, and the national regulatory framework are some of the most important factors that encourage the research of a more efficient energy generation process for internal combustion engines [31].

The main contribution of this paper is to present the comparative analysis results of some energetic and exergetic performance indicators of a 2-MW natural gas engine integrated with waste heat recovery systems based on SORC, RORC, and DORC configurations with different organic working fluids such as toluene, cyclohexane, and acetone. The maximum net output power of the bottoming ORC cycle was studied for differents engine load percentages, which implied studying the exhaust gas thermodynamic properties and engine thermal performance in detail. Also, parametric studies are developed to identify the influence of evaporation pressure on exergy destruction for each component, and performance indicator. These result help obtain the best ORC operational condition and configuration in terms of rational use of energy and environment preservation by increasing the overall energy and exergy efficiency of the Jenbacher JMS 612 GS-N. L natural gas engine of 2 MW with the bottoming ORC cycle.

2. Methodology

2.1. Description of the System

This section presents a general description of the 2-MW natural gas engine and each ORC configuration and operation. In this study, a Jenbacher JMS 612 GS-N. L engine (Figure 1) is considered to evaluate the ORC–WHR systems. The main design and operation engine parameters are summarized in Table 1. Fuel composition and air–fuel mixture supply conditions for the studied plastic production plant are presented in Table 2. The air–fuel mixture conditions are set to obtain an optimum flammable gas–air mixture and an exhaust gas temperature of 420–460 °C after the turbocharger expander.

Figure 1. Jenbacher JMS-612 GS-N. L natural gas engine.

Table 1. Main design and operation parameters of Jenbacher JMS 612 GS-N. L engine.

Description	Value	Units
Cylinder capacity	74.852	L
Compression ratio	10.5	–
Number of cylinders (In V–60°)	12	–
Stroke length	220	mm
Diameter in chamber	190	mm
Maximum torque	60.66	kN·m
Power at nominal speed	1820	kW
Nominal speed	1500	rpm
Ignition system	Spark ignition	–
Minimum load capacity	1000	kW
Maximum load capacity	1982	kW
λ at minimum load	1.79	–
λ at maximum load	1.97	–
Exhaust gases O_2 concentration	9.45–10.52	% volume
Exhaust gases CO concentration	588–731	mg/m^3
Exhaust gases NO_x concentration	461–468	mg/m^3
Exhaust gases NO_2 concentration	317–368	mg/m^3
Exhaust gases NO concentration	65–95	mg/m^3

Table 2. Fuel composition and supply conditions for the Jenbacher JMS 2 GS-N. L engine.

Description	Value	Units
Methane (CH_4)	97.97	%
Nitrogen (N_2)	1.50	%
Ethane (C_2H_6)	0.25	%
Carbon dioxide (CO_2)	0.16	%
Fuel–air mixture supply pressure	1.15–1.25	bar
Uncorrected volumetric ratio	110–140	L/s

2.2. Description of the SORC, RORC, and DORC

The studied ORC configurations are presented in Figure 2 and are based on the previous work of Fontalvo et al. [28]. The SORC physical structure is shown in Figure 2a. SORC operation is as follows: an air–fuel mixture (stream 1) is compressed to mixing conditions (stream 6) and supplied to the combustion chamber. The exhaust gases at the manifold outlet (stream 9) are expanded at the turbocharger expansion stage (stream 10) before they transfer heat to the thermal oil circuit at the shell and tube heat exchanger (ITC 1), and they are finally sent to the atmosphere (stream 11). In the thermal oil circuit, the oil that receives heat from the exhaust gasses (stream 1 AT) is used to preheat (zone 1), evaporate (zone 2), and superheat (zone 3) the ORC working fluid in the evaporation unit (ITC 2) before it is pumped (streams 3AT) and sent to the shell and tube heat exchanger (ITC 1). In the SORC, the high-pressure organic working fluid (1 ORC) is expanded at the ORC turbine (T 1) to the SORC lowest pressure (2 ORC) before it is cooled down and condensed (ITC 3). Then, the fluid is pumped to the ORC highest pressure (4 ORC) and sent back to the preheating, evaporating, and superheating zones to complete the cycle.

The ORC configuration with internal heat recovery (RORC) is shown in Figure 2b. This configuration is based on the SORC presented above, but an internal recovery unit (RC) is included in the ORC system to improve conversion efficiency by using the turbine outlet stream (2 ORC stream) to preheat the fluid in the pump outlet stream (5 ORC) before it enters the preheating zone in the evaporation heat exchanger (ITC 2).

The DORC configuration is shown in Figure 2c. This configuration uses two evaporating pressures and requires two evaporation units (ITC 2, ITC 4), two pumps (B 2, B 3), two turbines (T 1, T 2), and one condenser (ITC 3). In the DORC configuration, the thermal oil that leaves the pump (stream 3 AT) is heated by the exhaust gases (stream 10) in the shell and tube heat exchanger (ITC 1) before it is sent to the high-pressure evaporation unit (ITC 4) where the thermal oil (stream 1 AT) is used to preheat, evaporate, and superheat the high-pressure ORC working fluid (stream 6 ORC). Then, the thermal oil leaving the high-pressure evaporation unit (stream 1-2 AT) is used to preheat, evaporate, and superheat the mid-pressure ORC fluid (stream 5″ ORC) in the mid-pressure evaporation unit (ITC 2). In the ORC system, the high-pressure working fluid (stream 6 ORC) enters the first turbine stage (T 2), where it is expanded to the system middle pressure and mixed with the fluid (stream 2 ORC), leaving the mid-pressure evaporation unit. The mixed fluid (stream 2″ ORC) enters the second turbine stage (T 1) where it is expanded to the system lowest pressure (stream 3 ORC) before it is cooled down and condensed (ITC 3). Then, the fluid is pumped to the ORC middle pressure (5 ORC) before it is split (streams 5′ ORC and 5″ ORC). Stream 5′ is pumped to the system highest pressure and sent back to the high evaporation unit, while stream 5″ is sent to the mid-pressure evaporation unit to complete the cycle. According to the literature, the DORC configuration increases the efficiency of the cycle by decreasing the thermal load dissipated to the environment [32].

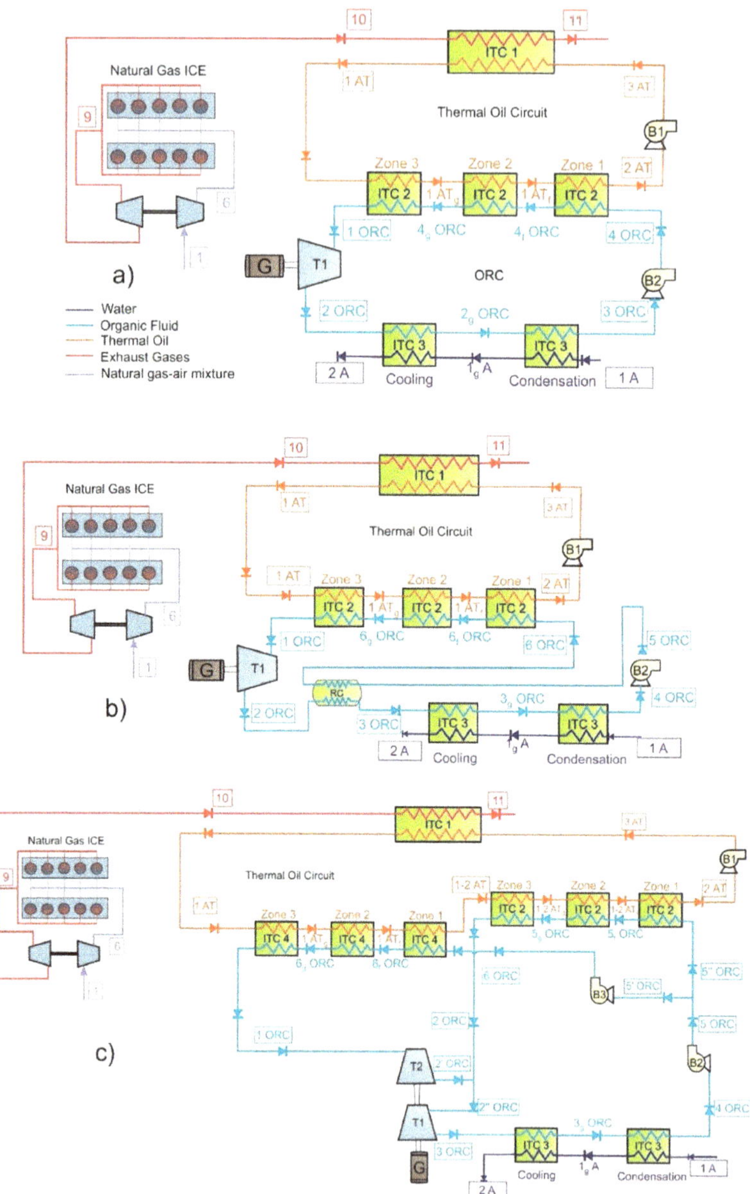

Figure 2. Physical structure of the waste heat recovery (WHR) system, (**a**) Simple organic Rankine cycle (SORC), (**b**) ORC with Recuperator (RORC), and (**c**) ORC with Double Pressure (DORC).

3. Thermodynamic Modeling

Energy and exergy analyses are conducted for the ORC–WHR configurations by applying the first and second law of thermodynamics to each component in the configurations. In addition, a definition of Fuel-Product is presented for each configuration component, as well as first and second law performance metrics.

3.1. Energy Analysis

Every single ORC–WHR system component is considered as a control volume. Mass and energy balances are applied to each component according to Equations (1) and (2), respectively.

$$\sum \dot{m}_{in} - \sum \dot{m}_{out} = 0 \tag{1}$$

$$\sum \dot{m}_{in}h_{in} - \sum \dot{m}_{out}h_{out} + \sum \dot{Q} + \sum \dot{W} = 0 \tag{2}$$

where $\dot{m}$ is the mass flow rate, h is the fluid specific enthalpy, and $\dot{Q}$. and $\dot{W}$ are the heat flow rate output and power inputs, respectively. Steady-state operation is assumed for each ORC component.

3.2. Exergy Analysis

The specific exergy is calculated by neglecting the variation of kinetic and potential energy, and it is calculated according to Equation (3):

$$ex = (h - h_0) - T_0(s - s_0) \tag{3}$$

where h_0 and s_0 are the reference state enthalpy and entropy, respectively, which are calculated at the reference conditions of $T_0 = 298.15$ K and $P_0 = 101.325$ kPa. The chemical exergy of the exhaust gases (stream 10) is calculated according to Equation (4) by considering a mixture of gases from the fuel combustion.

$$ex_G^{ch} = \sum_{i=1}^{n} X_i ex^{ch_i} + RT_0 \sum_{i=1}^{n} X_i ln X_i, . \tag{4}$$

where X_i is the molar fraction and ex^{ch_i} is the exergy per mol unit for each gas.

The exergy balance is applied to each ORC–WHR system component by means of Equation (5) [33]:

$$\sum \dot{m}_{in}ex_{in} - \sum \dot{m}_{out}ex_{out} + \dot{Q}\left(1 - \frac{T_0}{T}\right) - \dot{W} - \dot{Ex}_D = 0 \tag{5}$$

where $\dot{m}_{in}ex_{in}$ is the fluid incoming exergy flow, $\dot{m}_{out}ex_{out}$ is the fluid outcoming exergy flow, and $\dot{Ex}_D$ is the exergy destruction.

Three performance metrics are used for the ORC–WHR systems: cycle thermal efficiency $(\eta_{I,\,c})$, heat recovery efficiency (ε_{hr}), and overall energy conversion efficiency $\left(\eta_{I,\,overall}\right)$ [34]. Cycle thermal efficiency is calculated by means of Equation (6), while heat recovery efficiency is calculated using Equation (7), and overall energy conversion efficiency is calculated as shown in Equation (8).

$$\eta_{I,\,C} = \frac{\dot{W}_{net}}{\dot{Q}_G} \tag{6}$$

$$\varepsilon_{hr} = \frac{\dot{Q}_G}{\dot{m}_{10}C_{P10}(T_{10} - T_0)} \tag{7}$$

$$\eta_{I,\,overall} = \eta_{I,\,C} \cdot \varepsilon_{hr} \tag{8}$$

Similarly, the absolute increase in thermal efficiency is calculated using Equation (9), which relates the ORC net power output and fuel energy. This indicator determines the improvement in the ICE performance by using a WHR–ORC recovery system.

$$\Delta\eta_{thermal} = \frac{\dot{W}_{net}}{\dot{m}_{fuel} \cdot LHV} \tag{9}$$

Due to the ORC power output, there is a lower brake-specific fuel consumption (BSFC), which is calculated by Equation (10) [15]. The absolute decrease in specific fuel consumption is determined by Equation (11), which represents the reduction in fuel consumption at any particular operating conditions of the power generation engine:

$$BSFC_{ORC-engine} = \frac{\dot{m}_{fuel}}{\dot{W}_{engine} + \dot{W}_{net}} \tag{10}$$

$$\Delta BSFC = \frac{\left| BSFC_{ORC-engine} - BSFC_{engine} \right|}{BSFC_{engine}} \cdot 100 \tag{11}$$

The exergy efficiency based on the second law of thermodynamics $(\eta_{II,ORC})$. is calculated as shown in Equation (12):

$$\eta_{II,ORC} = \frac{\dot{Ex}_{produced}}{\dot{Ex}_{supplied}} \tag{12}$$

where $\dot{Ex}_{supplied}$ is the exergy supplied to the system and $\dot{Ex}_{produced}$ is the exergy recovered by the system. Exergy efficiency can be expressed as a function of the destroyed exergy $\dot{Ex}_D$ by means of Equation (13):

$$\eta_{II,ORC} = 1 - \frac{\dot{Ex}_D}{\dot{Ex}_{supplied}} \tag{13}$$

Simulating each WHR–ORC configuration, the following assumptions were considered:

- Pressure drops in pipelines are neglected.
- Pressure drops in heat exchangers are calculated as a function of the equipment geometry and the hydraulic flow characteristics.
- All the WHR–ORC components of the cycle are thermally insulated.
- The thermal oil circuit absorbs temperature variations in exhaust gases to obtain steady-state operation in each ORC configuration.

A simulation program with the energy and exergy analyses was written in MATLAB R2018b® [35], and the thermodynamic properties were calculated by using REFPROP 9.0® [36]. The detailed equations of energy balances applied to each configuration are shown in Appendix A, Table A1.

A specific engine operating condition is selected (see Table 3) to make it possible to compare each of the simulation results for each configuration. Values in Table 3 are also selected because it represents the system operation in off-grid mode. The engine performance indicators for the selected base conditions are shown in Table 4, which are expected to be evaluated with each configuration. The design parameters considered for the WHR–ORC simulations are shown in Table 5. In addition, the input–output structure for the components of the proposed WHR–ORC systems is shown as in Table 6, as the exergy losses must be differentiated from the exergy destroyed in each configuration [37].

Table 3. Parameters considered for internal combustion engines (ICE) simulation.

Parameter	Value	Units
Gas flow	120	L/min
λ	1.784	-
$\dot{n}_{rev}$	1482	rpm
Gas pressure	1163.6	mbar
Throttle valve	80.0	%
Turbo bypass valve	9.1	%
Gas temperature	389	°C
Engine coolant temperature	63.9	°C

Table 4. Performance indicators for ICE.

Performance Indicators	Value	Units
Mechanical engine power	1758.77	kW
Effective engine efficiency	38.59	%
Heat recovery efficiency	40.78	%
Heat removed from exhaust gases	514.85	kW
Specific engine fuel consumption	177.65	g/kWh

Table 5. Parameters considered for proposed configurations (S = SORC, R = RORC, and DP = DORC).

Configuration	Parameter	Value	Units	Reference
S/R/DP	Isentropic efficiency turbines	80	%	[38]
S/R/DP	Isentropic efficiency pumps	75	%	[38]
S/R/DP	Cooling water temperature (T1A)	50	°C	
S/R/DP	Pinch Point condenser (ITC3)	15	°C	
S/R	Pressure Ratio B1	2.5		
DP	Pressure Ratio B1	11.09		
S/R/DP	Pinch Point evaporators (ITC2) (ITC4)	35	°C	
R	Recovery Effectiveness (RC)	85	%	[38]
DP	Pressure Ratio B2	20		
DP	Pressure Ratio B3	9		
S/R	Pressure Ratio B2	30		

Table 6. Fuel-Product definition for each configuration.

Component	Different Configurations of Waste Heat Recovery Systems Using ORC								
	SORC			RORC			DORC		
	Fuel	Product	Lost	Fuel	Product	Lost	Fuel	Product	Lost
I1	$\dot{E}_{x10}$	$\dot{E}_{x1AT} - \dot{E}_{x3AT}$	$\dot{E}_{x11}$	$\dot{E}_{10}$	$\dot{E}_{1AT} - \dot{E}_{3AT}$	$\dot{E}_{11}$	$\dot{E}_{x10}$	$\dot{E}_{x1AT} - \dot{E}_{x3AT}$	$\dot{E}_{x11}$
B1	$\dot{W}_{B1}$	$\dot{E}_{x3AT} - \dot{E}_{x2AT}$	-	$\dot{W}_{B1}$	$\dot{E}_{x3AT} - \dot{E}_{x2AT}$	-	$\dot{W}_{B1}$	$\dot{E}_{x3AT} - \dot{E}_{x2AT}$	-
ITC2	$\dot{E}_{x1AT} - \dot{E}_{x2AT}$	$\dot{E}_{x1ORC} - \dot{E}_{x4ORC}$	-	$\dot{E}_{x1AT} - \dot{E}_{x2AT}$	$\dot{E}_{x1ORC} - \dot{E}_{x6ORC}$	-	$\dot{E}_{x1-2AT} - \dot{E}_{x2AT}$	$\dot{E}_{x2ORC} - \dot{E}_{x5ORC}$	-
T1	$\dot{E}_{x1ORC} - \dot{E}_{x2ORC}$	$\dot{W}_{T1}$	-	$\dot{E}_{x1ORC} - \dot{E}_{x2ORC}$	$\dot{W}_{T1}$	-	$\dot{E}_{x2''ORC} - \dot{E}_{x3}$	$\dot{W}_{T1}$	-
ITC3	-		$\dot{E}_{2A}$	-	-	$\dot{E}_{2A}$	-	-	$\dot{E}_{2A}$
B2	$\dot{W}_{B2}$	$\dot{E}_{x4ORC} - \dot{E}_{x3ORC}$	-	$\dot{W}_{B2}$	$\dot{E}_{x5ORC} - \dot{E}_{x4ORC}$	-	$\dot{W}_{B2}$	$\dot{E}_{x5ORC} - \dot{E}_{x4ORC}$	-
RC	-	-	-	$\dot{E}_{x2ORC} - \dot{E}_{x3ORC}$	$\dot{E}_{x6ORC} - \dot{E}_{x5ORC}$	-	-	-	-
T2	-	-	-	-	-	-	$\dot{E}_{x1ORC} - \dot{E}_{x2'ORC}$	$\dot{W}_{T2}$	-
B3	-	-	-	-	-	-	$\dot{W}_{B3}$	$\dot{E}_{x6ORC} - \dot{E}_{x5ORC}$	-
ITC4	-	-	-	-	-	-	$\dot{E}_{x1AT} - \dot{E}_{x1-2AT}$	$\dot{E}_{x1ORC} - \dot{E}_{x6ORC}$	-

3.3. Validation

The SORC simulation model is validated by conducting a comparative analysis using previously reported results from Vaja and Gambarotta [9] and Tian et al. [25]. The main parameters considered in both references are shown in Table 7. Two working fluids are considered: R-11 and R-134a, and ORC thermal efficiencies are determined as a function of the turbine inlet pressure.

Table 7. Data used from the system for model validation.

Cycle	η_P	η_T	T_{source} (°C)	F (kg/s)	T_{source} (°C)	Pinch Point (°C)	P_{vap} (MPa)
SORC	0.8	0.7	250	2.737	35	30	0.8–5.5
RORC	0.95	0.89	165	84.36	15	10	0.31

The validation results in Figure 3 evidence that the SORC model is accurate enough when it is compared to previous results from references [9] and [25]. When results are compared to Vaja and Gambarotta [9], the error is below 3% for R-11, below 6% for R-134a when the turbine inlet pressure is up to 1 MPa, and less than 3% when the turbine inlet pressure is above 2 MPa for R-134a. The pressure range in this study is set as 2.5–3 MPa; therefore, it can be observed that the relative error in Figure 3 is below 1% compared to that of Huan Tian et al. [25], which guarantees the accuracy of the results.

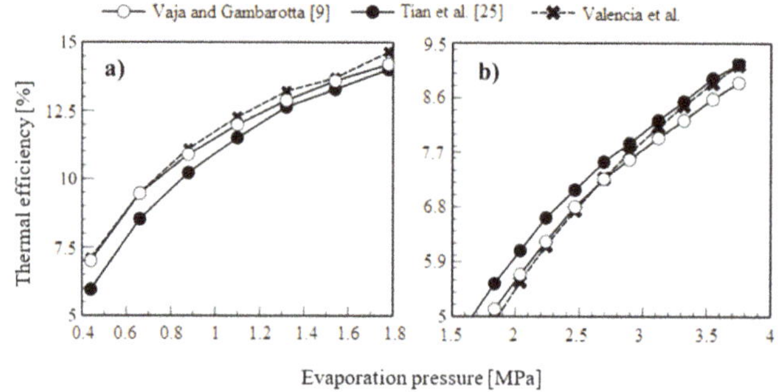

Figure 3. Thermal efficiency of the ORC as a function of the turbine inlet pressure for (**a**) R-11, (**b**) R-134a.

For RORC model validation, the results of this system for a geothermal application were taken from Emam et al. [39] and Zare et al. [27]. The parameters considered for both investigations are shown in Table 7. The following considerations were assumed to perform the comparative analysis of the RORC:

- The processes and subsystems are assumed to be in steady state.
- All devices were considered in adiabatic conditions.
- Pressure drops in ORC devices and pipelines are neglected.
- The reference temperature for exergy calculations is 288 K.

Validation results for RORC are shown as in Table 8. A good agreement can be seen between this study and the previously published ORC performances from Emam et al. [39] and Zare et al. [27]. For isobutane, error ranges between 0.62–0.73% for thermal efficiency. On the other hand, exergy efficiency relative error is between 0.18–0.35%.

Table 8. Validation of the proposed model for RORC.

Parameters	Valencia et al.	Emam et al. [39]	Zare et al. [27]
$T_{1\,AT}$ (°C)	165	165	165
$\dot{m}_{1\,AT}$ (kg/s)	84.36	84.36	82.16
$\dot{m}_{1\,ORC}$ (kg/s)	75.22	78.06	76.09
$A_{ITC\,2}$ $\left(\text{m}^2\right)$	394.21	399.30	390.60
$A_{ITC\,3}$ $\left(\text{m}^2\right)$	809.52	810.10	808.70
A_{RC} $\left(\text{m}^2\right)$	124.58	124.80	124.20
η_{th} (%)	16.25	16.37	16.15
η_{exe} (%)	48.71	48.80	48.54

4. Results and Discussions

From the simulation results obtained, it is possible to determine and calculate the stream and fluid properties for the proposed SORC, RORC, and DORC configurations. Detailed simulation results are shown in Appendix B, Table A2 for the case of SORC. A summary of the main results obtained for the engine WHR system is shown in Table 9.

Table 9. Parameters of an integrated recovery system with engine and ORC. BSFC: brake-specific fuel consumption.

Parameters	SORC	RORC	DORC	Units
Thermal efficiency engine–ORC	40.45	41.25	40.72	%
Increased thermal efficiency	1.86	2.66	2.13	%
Thermal efficiency ORC	16.40	23.51	18.74	%
Global energy conversion efficiency	6.68	9.59	7.66	%
Global exergetic efficiency	34.5	49.47	39.43	%
BSFC engine–ORC	169.5	166.21	168.42	g/kWh

Results from the exergy analysis are presented in Table 10. Results in this table are calculated through the input–output definition in every component of the SORC configuration. The exergy destruction fraction in each component of the cycle is calculated from the results of the input–output definition and the exergy destruction percentages are presented in Figure 4a.

Table 10. Exergy analysis results for each component of the heat recovery system with SORC.

Component	Input (kW)	Product (kW)	$\dot{E}x_D$ (kW)	Lost (kW)
ITC1	541.202051	202.794262	41.9535673	296.454222
B1	0.37472727	0.05848531	0.31624196	–
ITC2	202.852748	166.340104	36.5126437	–
T1	99.4808146	85.5899807	13.8908338	–
ITC3	–	–	36.0581887	66.5877282
B2	0.75619324	0.58683347	0.16935977	–

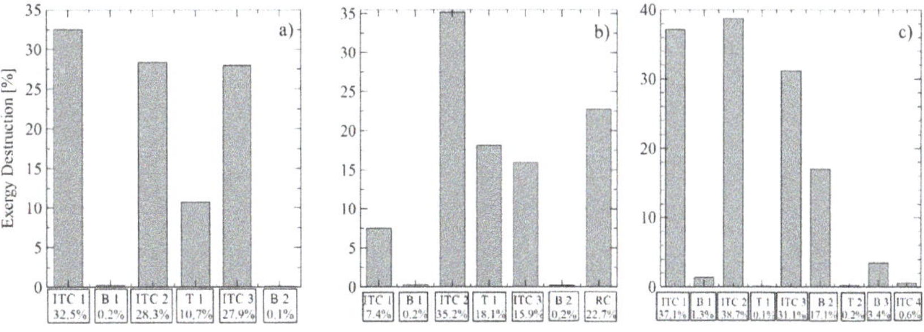

Figure 4. Exergy destruction by heat recovery system component (**a**) SORC, (**b**) RORC, and (**c**) DORC.

It is observed that the shell and tube heat exchanger (ITC 1) has the highest exergy destruction by achieving 32.5% of the cycle's exergy destruction, followed by the evaporation unit (ITC 2) with 28.3%, and the condenser (ITC 3) with 27.9%. On the other hand, pumps show the lowest exergy destruction contribution. Due to operational restrictions on the turbine inlet organic fluid temperature, it is not possible to reduce the evaporation unit minimum temperature to obtain less generation of entropy and less exergy destruction in this equipment. Nevertheless, these temperature differences in heat exchangers can be optimized for better performance.

The use of isentropic or dry fluids in the RORC configuration allows obtaining superheated and relatively high-temperature conditions at the turbine outlet (2 ORC stream) [40]. This is the case of organic fluid toluene in RORC, as shown in Appendix B, Table A3. Therefore, the recovery heat exchanger (RC) uses this energy at the turbine outlet to preheat the stream leaving the pump (6 ORC stream) [41]. This preheating strategy [42] increases the thermal efficiency by 2.6%, while in SORC, the increase is 1.8%. Thus, the whole engine–WHR–RORC configuration shows a specific fuel consumption of 166.21 g/kWh, which is 2% less than the engine–WHR–SORC specific fuel consumption as shown in Table 10 by using the same fluid and operation conditions.

It can be seen that the ORC performance increases when the recovery heat exchanger (RC) is included, as it increases the turbine power output for the same heat input from the exhaust gases. The efficiency improvement is strongly related to fluid properties, especially to specific heat [43]. The results confirm that the recovery system does not affect the turbine power output or pump power consumption, but it modifies heat transfer in both the evaporation unit and the condenser [44], which have the highest exergy destruction, as shown in Table 11.

Table 11. Exergy analysis results for each component of the heat recovery system with RORC. RC: internal recovery unit.

Component	Input (kW)	Product (kW)	$\dot{Ex}_D$ (kW)	Lost (kW)
ITC1	541.202051	237.549898	7.197931	296.454222
B1	0.36222219	0.09598314	0.26623905	–
ITC2	237.645881	203.818309	33.827572	–
T1	139.797975	122.396844	17.40113113	–
ITC3	–	–	15.3197	71.5691491
B2	0.94533621	0.7336153	0.21172091	–
RC	86.4902041	64.620999	21.8692051	–

The exergy destruction contribution in each cycle component is presented in Figure 4b. From this figure, it can be seen that the exergy destruction is 33.82 kW (35.2%) in the evaporator, followed by the recuperator with 21.86 kW (22.8%). Also, it can be seen that pump contribution to exergy destruction is only 0.5%.

For the DORC configuration, there are two different evaporating pressures; however, the 2 ORC, 2' ORC, and 2" ORC streams are mixed at the same pressure, and then enter the turbine (T1). In addition, at the low-pressure pump outlet (B2), the states 5 ORC, 5' ORC, and 5" ORC have similar thermodynamic properties with different mass flow rates, as shown in Appendix B, in Table A4.

Table 9 presents the performance metrics of engine–WHR–DORC. It can be observed that this configuration only achieves an overall energy conversion efficiency of 6.68% and an overall exergy efficiency of 34.37% at the test conditions. According to Guzovi [45], these results are strongly related to the selected pressure ratio values, and this configuration presents better exergy efficiency and generates more power output than the SORC and the RORC configurations. In this configuration, the high-pressure evaporator (ITC 4) along with the shell and tube heat exchanger (ITC1) show the highest exergy destruction, as shown in Table 12, which can be explained by the temperature differences between the thermal oil and the organic fluid.

Table 12. Exergy analysis results for each component of the heat recovery system with DORC.

Component	Input (kW)	Product (kW)	$\dot{Ex}_D$ (kW)	Lost (kW)
ITC1	541.20	203.57	31.15	296.45
B1	1.66	0.26	1.40	–
ITC2	113.18	101.93	11.26	–
T1	95.39	79.75	15.63	–
ITC3	–	–	31.01	6.00
B2	0.63	0.49	0.14	–
T2	10.27	8.77	1.49	–
B3	2.09	1.63	0.47	–
ITC4	90.64	52.66	37.99	–

The component exergy destruction contribution to the WHR–DORC system is shown in Figure 4c. From this figure, the evaporators and condensers' contribution is 86.4% of the total exergy destruction. As the organic fluid is evaporated in the heat exchangers, a closer match can be maintained between the thermal oil cooling temperature profile and the ORC working fluid temperature profile, which reduces exergy losses.

4.1. Sensitivity Analysis

4.1.1. Effect of Evaporation Pressure

Considering the relevance of the evaporation pressure on the operation of these systems, a comparative analysis of the net power, an absolute increase of thermal efficiency, and absolute decrease of the specific fuel consumption is presented in Figure 5 for each engine–WHR–ORC configuration. For the DORC, high evaporation pressure is reached by changing the pressure ratio at the B3 pump, while the pressure ratio at the B2 pump is kept fixed at three for toluene, 10 for acetone, and 1.5 for cyclohexane.

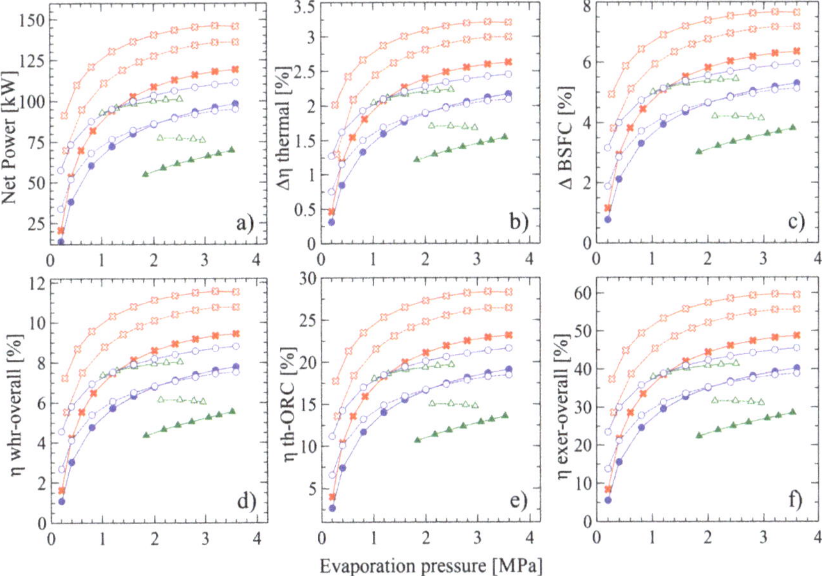

Figure 5. Performance of SORC, RORC, and DORC configurations with different organic fluids, (**a**) Net power, (**b**) Absolute increase in thermal efficiency, (**c**) Absolute decrease in specific fuel consumption, (**d**) Global energy conversion efficiency, (**e**) ORC thermal efficiency, and (**f**) Global exergetic efficiency.

It is observed that the net power output is strongly related to the evaporation pressure, achieving the most profitable results for the toluene–RORC at an evaporation pressure of 3.4 MPa: 146.25 kW of net power, which is up to 31.9% more power than the toluene–SORC net power at the same evaporation pressure. It is not possible to achieve the same evaporating pressure in the DORC by changing the B3 pump pressure ratio, because the fluid temperature would be so high that the heat transfer in the evaporation unit will be reversed.

When comparing the absolute increase in thermal efficiency for the three ORC configurations, Figure 5b shows that the evaporating pressure increases thermal efficiency up to an upper limit [46,47]. For toluene at 3.4 MPa, the SORC increases engine efficiency by 2.44%, while the RORC increases it by 3.22%. Working fluids such as acetone and cyclohexane in the SORC achieve an absolute thermal efficiency increase of 2.15% and 2.09%, respectively. Therefore, toluene stands out as the working fluid that presents the best results performance. Furthermore, as the main objective of this research is to increase the overall conversion efficiency by generating additional power, a RORC is the best option, regardless of the working fluid. This is supported by Figure 5, where the highest specific fuel consumption reduction is 5.92% for the toluene–SORC combination and 7.67% for the toluene–RORC combination, while the lowest specific fuel consumption reduction is achieved by the acetone–SORC cycle, which is 0.78% at 0.2 MPa. However, it is important to mention that as the heat in the evaporator increases, the size of the evaporation unit in the RORC will increase the purchase equipment cost and reduce the revenue of the system.

By evaluating the overall energy conversion efficiency, as shown in Figure 5d, the toluene–SORC and the cyclohexane–SORC combinations achieve 8.78% and 7.54%, respectively, with 3.4 MPa. For the RORC configuration, acetone achieved the lowest efficiency (4.91%), which confirms that toluene is the potential fluid to be used in SORC and RORC.

The results also show that the toluene–RORC combination increases thermal efficiency from 21.53% to 28.41%, and increases global exergy efficiency from 45.29% to 59.76%, which confirms that toluene stands out among the studied fluids, at relatively low pressures.

The maximum net power output in the toluene–SORC combination represents 8.31% of the stationary motor-generating capacity at nominal speed. In addition, it is observed that toluene–SORC working at evaporation pressures between 2–3 MPa increases the net power output by 5.49% and up to 109.3 kW. The absolute increase in thermal efficiency goes up to 5.26%. Finally, the specific fuel consumption reduction increases by 5.21%. On the other hand, the increases for these variables in the RORC are 3.76%, 3.75%, and 3.47%, which allows us to conclude that toluene–SORC is the combination that stands out when considering this variable.

4.1.2. Analysis of the Influence of Evaporation Pressure on the Destruction of Exergy Configurations

To perform a comparative analysis of the exergetic performance for each ORC configuration, the fractions of exergy destruction in each component of the cycle were calculated with the three proposed working fluids. The exergy destruction fractions for each ORC configuration are shown in Figure 6, at different evaporating pressures for acetone, cyclohexane, and toluene.

The results demonstrate that the exergy fractions for the DORC cycle are higher than those for the SORC cycle, while the RORC cycle has the lowest values of total exergy loss.

After the evaluation of the average decrease in the percentage points of the three configurations studied to the engine operating condition, it was demonstrated that the DORC cycle working with toluene presents an average of 74.02% total exergy destruction, which is more significant than the exergy destruction in the RORC configuration. These results are due to the DORC configurations leading to higher exergy destruction by having more components. However, in an operational range of evaporating pressure, the dual ORC exceeds the destroyed exergy of the SORC by 9%. Consequently, the evaporating pressure must be determined for each specific case to achieve the highest turbine output power, and thus the greater exergetic efficiency of the system.

The DORC configuration presents the highest values of exergy destruction fractions operating with acetone within the temperature range of the source studied, while the SORC configuration presents the lowest values of the exergy destruction fraction. As a result, the SORC configuration delivers more power with the least heat rejection in the condenser; additionally, for the same power delivered, this configuration requires less heat from the heat source. The results also show that the evaporating pressure in the different ORC configurations has a significant effect on the total dimensionless exergy losses, with a minimum value that is within the studied range.

The maximum values of exergy losses occurred in the condenser for every configuration, with a maximum value of 88.26% for the SORC at 0.803 MPa. As a result, the exergy destruction fractions in the condenser is inversely proportional to the evaporating pressure as it approaches 2 MPa. Due to the significant increase in exergy losses, the recuperator acquires importance in the evaporator and turbine cycle in the RORC, while in the dual cycle, the low evaporator provides exergy destruction as the pressure increases.

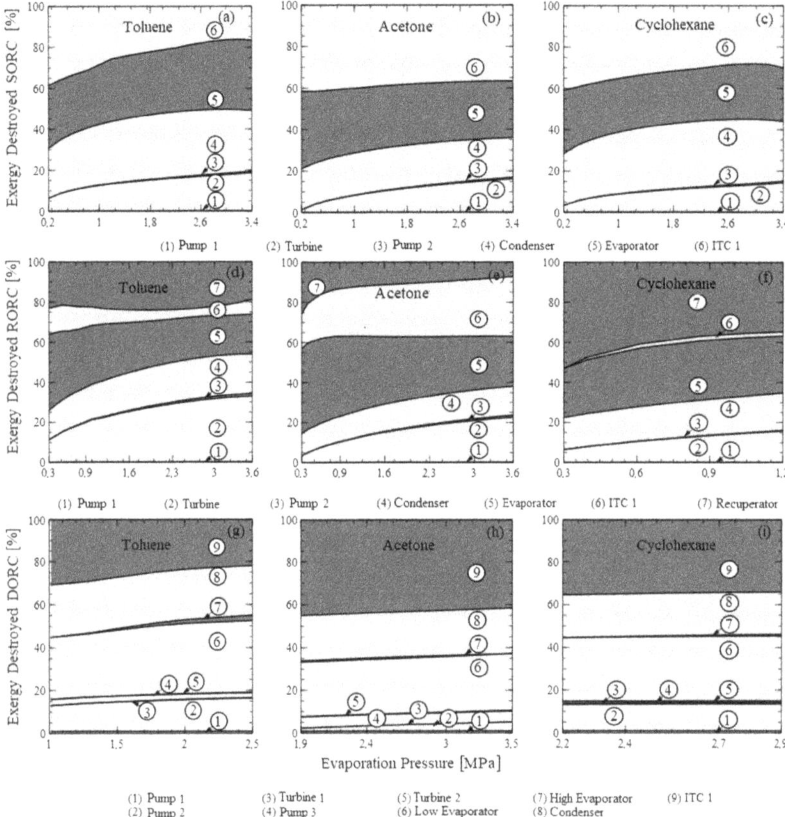

Figure 6. Exergy destruction by component as a function of evaporating pressure for (a–c) configuration SORC, (d–f) configuration RORC, and (g–i) configuration DORC.

4.1.3. Analysis of the Influence of Engine Load on Energy Performance

This section analyzes the effect of engine load on the performance of the heat recovery system. The results presented above were obtained based on the typical operating condition of a natural gas engine. The engine power control system adjusts internal engine variables such as mixture pressure, temperature, and mixture recirculation percentage to provide high efficiency in operations with partial engine loads. The global energy indicators were selected as study variables, and the results of the three configurations under study for an evaporating pressure of 675.8 kPa working with toluene are shown as in Figure 7. For safety reasons, all the possible operating points of the proposed configurations at different engine loads guarantee that toluene vaporizes completely at the evaporator outlet in order to prevent corrosion of the liquid in the expander. Moreover, the engine exhaust gas temperature at the evaporator outlet (stream 11) must be higher than the acid spray temperature (200 °C) to prevent the acid corrosion of the exhaust.

The thermal efficiency presents a directly proportional relationship with respect to the engine load increases, while the overall energy conversion efficiency presents an inversely proportional relationship. The maximum net output power obtained for configurations at engine load percentages is SORC (89.4 kW—97.9%), RORC (124.5 kW—97.9%), and DORC (86.29 kW—91.81%). However, in an engine operating interval, the thermal efficiency increase as the RORC configuration increases first and then decreases, presenting a maximum performance with 82.68% engine load.

These results are due to the engine load being directly related to the flow in the exhaust gases and the energy loss in the recuperator since the evaporation pressure and the temperatures of the thermal coupling oil have been restricted. As the operating load increases, there is an increase in the evaporation temperature of the organic fluid. Therefore, the power increases, which is the main factor that affects the thermal and exergetic efficiency. However, the isentropic efficiency of the turbine decreases slightly as a consequence of the increase in the temperature of the thermal oil, causing a decrease in the indicators at high engine loads. The direct relation between the net power and the engine load is also due to the increase of the thermal oil inlet temperature to the evaporator, which leads to an increase in the mass flow of toluene, and the enthalpy difference between the pump and the turbine; however, this causes a stronger effect in the turbine.

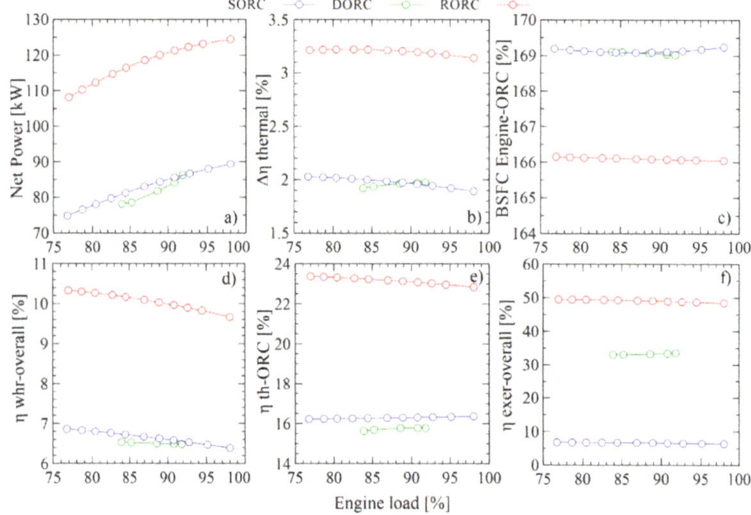

Figure 7. Performance of SORC, RORC, and DORC configurations with different engine loads: (**a**) Net power, (**b**) Absolute increase in thermal efficiency, (**c**) Specific fuel consumption of engine–ORC, (**d**) Overall energy conversion efficiency, (**e**) ORC thermal efficiency, and (**f**) Overall exergetic efficiency.

5. Conclusions

This study performed an energy and exergey analysis of three ORC systems for WHR from the exhaust gases of a 2-MW Jenbacher JMS 612 GS-N. L natural gas as a stationary engine at a plastic industry located at Barranquilla, Colombia. In particular, a validated thermodynamic model was employed to determine the net power output, fuel consumption, and thermal and exergy efficiency of the proposed engine–WHR–configurations based on the mean variables of the system. The study involved the calculation of energy and exergy performance metrics for the proposed systems to determine the improvement of the stationary engine overall energy conversion.

To improve the system performance, the irreversibilities and exergy destruction of the components must be reduced. The exergy destruction of all the elements in the proposed configurations is lower than the thermal oil pump (B1). These values suggest that reducing the heat transfer area in the evaporator, recuperator, and condenser may provide a favorable solution, especially in the DORC configuration. However, it is important to point out that these plate heat exchangers are manufactured from specialized materials, which contributes to the total purchase equipment cost. Also, the operation of these components has a significant effect on the total exergy destruction and thermal efficiency as a result of the pinch point temperature. Therefore, increasing the heat exchangers area increases the cost of generating electricity.

The thermal oil pump is the component with the lowest efficiency: SORC (16%), RORC (26%), and DORC (17%). However, among the heat exchange equipment, the shell and tube heat exchanger (ITC1) has the lowest exergy efficiency: SORC (37%), RORC (44%), and DORC (38%), because the organic fluid cannot reach the engine exhaust gas temperature levels, as heat transfer irreversibilities will increase and due to thermal stability conditions. Hence, these alternatives present better results for medium and low-temperature exhaust gases. Heat exchanger ITC1 has the highest contribution to exergy destruction for SORC and DORC, while the evaporation unit (ITC2) shows the highest contribution for RORC. Therefore, special effort should be focused to reduce the exergy destruction in these components. Based on commercial information about the geometric characteristics of plate heat exchangers and shell and tube heat exchangers, optimal sizes of these components should be determined to reduce cost and increase performance.

The highest specific fuel consumption reduction was 5.92% for the toluene–SORC combination, and 7.67% for the toluene–RORC combination. Also, the thermal efficiency (28.41%) and global exergy efficiency (59.76%) confirm that toluene–RORC assembly is the best alternative for this natural gas engine.

Author Contributions: Conceptualization: G.V.; Methodology: G.V., J.D. and A.F.; Software: G.V., A.F. and Y.C.; Validation: G.V., A.F. and Y.C.; Formal Analysis: G.V., J.D. and C.I.; Investigation: G.V.; Resources: G.V. and J.D.; Writing-Original Draft Preparation: G.V.,A.F; Writing-Review & Editing: J.D. and C.I.; Funding Acquisition: G.V.

Funding: This work was supported by the Universidad del Atlántico through the "PRIMERA CONVOCATORIA INTERNA PARA APOYO AL DESARROLLO DE TRABAJOS DE GRADO EN INVESTIGACIÓN FORMATIVA—NIVEL PREGRADO Y POSGRADO 2018", Universidad Pontificia Bolivariana and the E2 Energía Eficiente S.A E.S. P company.

Conflicts of Interest: The authors declare no conflict of interest.

Abbreviations

The following abbreviations are used in this manuscript:

BSFC	Brake-specific fuel consumption
DORC	Double pressure organic Rankine cycle
GHG	Greenhouse gases
GWP	Global warming potential
ICE	Internal combustion engine
ORC	Organic Rankine cycle
RORC	Recuperator organic Rankine cycle
SORC	Simple organic Rankine Cycle
WHR	Waste heat recovery

Nomenclature

C_p	Specific heat at constant pressure (J/kg K)
E	Energy (J)
ex	Specific exergy (kJ/kg)
h	Specific enthalpy (kJ/kg)
LHV	Heating power (kJ/kg)
m	Mass (kg)
$\dot{m}$	Mass flow rate (kg/s)
Q	Heat (J)
R	Universal gas constant (atm L/mol K).
rpm	Rotational engine speed (rpm)
T	Temperature (K)
t	Time (s)
$\dot{W}$	Power (kW)
X_i	Molar gas fraction

Greek Letters

$\eta_{I,c}$	Thermal efficiency of the cycle
$\eta_{I,overall}$	Overall energy conversion efficiency
$\eta_{II,ORC}$	Exergetic efficiency
ε_{hr}	Heat recovery efficiency

Subscripts

D	Destroyed
in	Input
out	Output
G	Gases
VC	Control volume
o	Reference condition

Appendix A

The energy balance applied to each component of the proposed configurations is presented in Table A1.

Table A1. Energy balances for the components of each configuration.

Component	Different Configurations of Residual Heat Recovery Systems Using ORC		
	SORC	**RORC**	**DORC**
Heat Exchanger 1 (ITC1)	$\dot{Q}_G = \dot{m}_{10} C_{P10}(T_{10} - T_{11}) = \dot{m}_{AT} C_{PAT}(T_{4AT} - T_{3AT})$ (A1)	(A1)	(A1)
Pump 1 (B1)	$\eta_{B1} = \frac{v_{2AT}(P_{3AT} - P_{2AT})}{h_{3AT} - h_{2AT}}$ (A2) $\dot{W}_{B1} = \dot{m}_{AT}(h_{3AT} - h_{2AT})$ (A3)	(A2)	(A2)
Heat Exchanger (ITC2)	Zone 1 (Preheating) $\dot{Q}_{Z1} = \dot{m}_{AT}(h_{ATf} - h_{2AT}) = \dot{m}_{ORC}(h_{4fORC} - h_{4ORC})$ (A4) Zone 2 (Evaporation) $\dot{Q}_{Z2} = \dot{m}_{AT}(h_{ATg} - h_{ATf}) = \dot{m}_{Z2}(h_{fgORC})$ (A5) Zone 3 (Overheating) $\dot{Q}_{Z3} = \dot{m}_{AT}(h_{1AT} - h_{ATg}) = \dot{m}_{ORC}(h_{1ORC} - h_{4gORC})$ (A6)	$\dot{Q}_{Z1} = \dot{m}_{AT}(h_{ATf} - h_{2AT} = \dot{m}_{ORC}(h_{6fORC} - h_{6ORC})$ (A13) Zone 2 (A5) $\dot{Q}_{Z3} = \dot{m}_{AT}(h_{1AT} - h_{ATg}) = \dot{m}_{ORC}(h_{1ORC} - h_{4gORC})$ (A14)	$\dot{Q}_{Z1} = \dot{m}_{AT}(h_{ATf} - h_{2AT}) = \dot{m}_{ORC}(h_{5fORC} - h_{5ORC})$ (A21) Zone 2 (A5) $\dot{Q}_{Z3} = \dot{m}_{AT}(h_{1-2AT} - h_{ATg}) = \dot{m}_{5ORC}(h_{2ORC} - h_{5gORC})$ (A22)
Turbine 1 (T1)	$\eta_{T1} = \frac{h_{1ORC} - h_{2ORC}}{h_{1ORC} - h_{2SORC}}$ (A7) $\dot{W}_{T1} = \dot{m}_{ORC}(h_{1ORC} - h_{2ORC})$ (A8)	(A7) (A8)	$\eta_{T1} = \frac{h_{2''ORC} - h_{3ORC}}{h_{2''ORC} - h_{3SORC}}$ (A29) $\dot{W}_{T1} = (\dot{m}_{5ORC} + \dot{m}_{6ORC})(h_{2''ORC} - h_{3ORC})$ (A30)
Heat Exchanger (ITC3)	Cooling $\dot{Q}_{ZE} = \dot{m}_{ORC}(h_{2ORC} - h_{2gORC}) = \dot{m}_{1A}(h_{2A} - h_{1gA})$ (A9) Condenser $\dot{Q}_{ZC} = \dot{m}_{ZC}(h_{2gORC} - h_{3ORC}) = \dot{m}_{1A}(h_{1gA} - h_{1A})$ (A10)	$\dot{Q}_{ZE} = \dot{m}_{ORC}(h_{3ORC} - h_{3gORC}) = \dot{m}_{1A}(h_{2A} - h_{1gA})$ (A15) $\dot{Q}_{ZC} = \dot{m}_{ZC}(h_{3gORC} - h_{4ORC}) = \dot{m}_{1A}(h_{1gA} - h_{1A})$ (A16)	(A15) (A16)
Pump 2 (B2)	$\eta_{B2} = \frac{v_{3ORC}(P_{4ORC} - P_{3ORC})}{h_{4ORC} - h_{3ORC}}$ (A11) $\dot{W}_{B2} = \dot{m}_{ORC}(h_{4ORC} - h_{3ORC})$ (A12)	$\eta_{B2} = \frac{v_{4ORC}(P_{5ORC} - P_{4ORC})}{h_{5ORC} - h_{4ORC}}$ $\dot{W}_{B2} = \dot{m}_{ORC}(h_{5ORC} - h_{4ORC})$ (A18)	$\eta_{B2} = \frac{v_{4ORC}(P_{5ORC} - P_{4ORC})}{h_{5ORC} - h_{4ORC}}$ (A27) $\dot{W}_{B2} = (\dot{m}_{5ORC} + \dot{m}_{6ORC})(h_{5ORC} - h_{4ORC})$ (A28)
RC	-	$\varepsilon_{RC} = \frac{T_{2ORC} - T_{3ORC}}{T_{2ORC} - T_{5ORC}}$ (A19) $\dot{Q}_{RC} = \dot{m}_{ORC}(h_{6ORC} - h_{5ORC}) = \dot{m}_{ORC}(h_{2ORC} - h_{3ORC})$ (A20)	-
Turbine 2 (T2)	-	-	$\eta_{T2} = \frac{h_{1ORC} - h_{2ORC}}{h_{1ORC} - h_{2'SORC}}$ (A25) $\dot{W}_{T2} = \dot{m}_{6ORC}(h_{1ORC} - h_{2'ORC})$ (A26)
Pump 3 (B3)	-	-	$\eta_{B3} = \frac{v_{5ORC}(P_{6ORC} - P_{5ORC})}{h_{6ORC} - h_{5ORC}}$ (A31) $\dot{W}_{B3} = \dot{m}_{6ORC}(h_{6ORC} - h_{5ORC})$ (A32)
Heat Exchanger (ITC4)	-	-	$\dot{Q}_{Z1} = \dot{m}_{AT}(h_{ATf} - h_{1-2AT}) = \dot{m}_{6ORC}(h_{6fORC} - h_{6ORC})$ (A23) Zone 2 (A5) $\dot{Q}_{Z3} = \dot{m}_{AT}(h_{1AT} - h_{1-ATg}) = \dot{m}_{6ORC}(h_{1ORC} - h_{6gORC})$ (A24)

Appendix B

The properties of the main currents of the heat recovery system proposed with SORC are presented in Table A2.

Table A2. Properties considered for SORC configuration.

Stream	Flow (kg/s)	P (kPa)	T (°C)	Enthalpy (kJ/kg)	Entropy(S-S0) (kJ/kg K)	Exergy (kW)
10	2.77	102.30	435.07	−1960.35	0.90	541.20
11	2.77	101.30	270.00	−2143.67	0.59	296.45
1 AT	1.64	101.43	307.84	461.66	0.94	208.75
1 ATg	1.64	91.42	246.29	324.52	0.73	106.76
1 ATf	1.64	81.01	178.30	183.24	0.47	29.12
2 AT	1.64	68.15	142.65	113.96	0.31	5.90
3 AT	1.64	170.38	142.77	114.19	0.31	5.96
1 ORC	0.72	675.85	272.84	633.29	1.80	169.27
2 ORC	0.72	22.53	202.37	513.72	1.87	69.79
2 gORC	0.72	22.53	65.00	301.64	1.34	31.18
3 ORC	0.72	22.53	65.00	−87.53	0.19	2.35
4 ORC	0.72	675.85	65.31	−86.47	0.19	2.93
4 fORC	0.72	675.85	194.20	181.72	0.86	50.17
4 gORC	0.72	675.85	194.20	477.95	1.50	124.68
1 A	13.32	101.30	50.00	209.42	0.27	35.20
1 gA	13.32	101.30	55.00	230.33	0.33	54.44
2 A	13.32	101.30	57.72	241.72	0.37	66.59

The properties of the main currents of the heat recovery system proposed with RORC are presented in Table A3.

Table A3. Properties considered for RORC configuration.

Stream	Flow (kg/s)	P (kPa)	T (°C)	Enthalpy (kJ/kg)	Entropy(S-S0) (kJ/kg K)	Exergy (kW)
10	2.77	102.30	435.07	−1960.35	0.90	541.20
11	2.77	101.30	270.00	−2143.67	0.59	296.45
1 AT	1.51	101.43	374.47	618.96	1.25	263.92
1 ATg	1.51	92.78	379.55	631.30	1.27	273.93
1 ATf	1.51	83.89	294.56	431.40	0.98	126.19
2 AT	1.51	68.15	209.28	246.22	0.65	26.27
3 AT	1.51	170.38	209.40	246.46	0.65	26.37
1 ORC	0.89	675.85	339.47	775.16	2.05	272.11
2 ORC	0.89	22.53	268.70	638.39	2.11	132.31
3 ORC	0.89	22.53	102.23	352.50	1.49	45.82
3 gORC	0.89	22.53	65.00	301.64	1.34	38.98
4 ORC	0.89	22.53	65.00	−87.53	0.19	2.93
5 ORC	0.89	675.85	65.31	−86.47	0.19	3.67
6 ORC	0.89	675.85	194.20	199.42	0.90	68.29
6 fORC	0.89	675.85	194.20	181.72	0.86	62.72
6 gORC	0.89	675.85	194.20	477.95	1.50	155.86
1 A	16.65	101.30	50.00	209.42	0.27	44.00
1 gA	16.65	101.30	55.00	230.33	0.33	68.06
2 A	16.65	101.30	55.65	233.06	0.34	71.57

The properties of the main currents of the heat recovery system proposed with DORC are presented in Table A4.

Table A4. Properties considered for DORC configuration.

Stream	Flow (kg/s)	T (°C)	P (kPa)	Enthalpy (kJ/kg)	Entropy (S-S0) (kJ/kg K)	Exergy (kW)
10	2.77	435.07	102.30	−1960.35	0.90	541.20
11	2.77	270.00	101.30	−2143.67	0.59	296.45
1 AT	1.62	316.58	686.83	481.81	0.99	215.40
1 ATg	1.62	296.48	676.43	435.75	0.92	178.88
1 ATf	1.62	245.76	666.37	323.37	0.74	99.29
1-2 AT	1.62	171.41	656.79	169.58	0.44	20.13
1-2 ATg	1.62	228.96	647.85	287.41	0.68	77.18
1-2 Atf	1.62	183.04	638.85	192.71	0.49	29.19
2 AT	1.62	150.98	630.14	129.89	0.35	7.53
3 AT	1.62	151.51	755.78	130.91	0.35	7.80
1A	8.40	50.00	101.30	209.42	0.27	27.60
1g A	8.40	55.00	101.30	230.33	0.33	42.69
2A	8.40	55.00	101.30	234.54	0.34	46.10
1 ORC	0.45	281.58	1351.69	635.64	1.75	159.64
2 ORC	0.62	136.41	450.56	53.17	0.57	13.75
2′ ORC	0.45	252.65	450.56	597.65	1.77	132.33
2″ ORC	1.08	173.03	450.56	368.93	1.28	139.83
3 ORC	1.08	65.00	22.53	298.79	1.34	46.85
3g ORC	1.08	65.00	22.53	301.64	1.34	47.17
4 ORC	1.08	65.00	22.53	−87.53	0.19	3.55
5 ORC	1.08	65.21	450.56	−86.84	0.19	4.13
5′ ORC	0.45	65.21	450.56	−86.84	0.19	2.40
5″ ORC	0.62	65.21	450.56	−86.84	0.19	1.74
5f ORC	0.62	173.04	450.56	132.93	0.76	24.32
5g ORC	0.62	173.04	450.56	447.79	1.46	70.24
6 ORC	0.45	65.64	1351.69	−85.38	0.20	3.10
6f ORC	0.45	235.75	1351.69	283.87	1.07	68.73
6g ORC	0.45	235.75	1351.69	535.81	1.56	132.70

References

1. Nawi, Z.M.; Kamarudin, S.; Abdullah, S.S.; Lam, S. The potential of exhaust waste heat recovery (WHR) from marine diesel engines via organic rankine cycle. *Energy* **2019**, *166*, 17–31. [CrossRef]
2. Alshammari, F.; Pesyridis, A.; Karvountzis-Kontakiotis, A.; Franchetti, B.; Pesmazoglou, Y. Experimental study of a small scale organic Rankine cycle waste heat recovery system for a heavy duty diesel engine with focus on the radial inflow turbine expander performance. *Appl. Energy* **2018**, *215*, 543–555. [CrossRef]
3. Liu, P.; Shu, G.; Tian, H.; Wang, X.; Yu, Z. Alkanes based two-stage expansion with interheating Organic Rankine cycle for multi-waste heat recovery of truck diesel engine. *Energy* **2018**, *147*, 337–350. [CrossRef]
4. Michos, C.N.; Lion, S.; Vlaskos, I.; Taccani, R. Analysis of the backpressure effect of an Organic Rankine Cycle (ORC) evaporator on the exhaust line of a turbocharged heavy duty diesel power generator for marine applications. *Energy Convers. Manag.* **2017**, *132*, 347–360. [CrossRef]
5. Patel, P.S.; Doyle, E.F. *Compounding the Truck Diesel Engine with an Organic Rankine-Cycle System*; SAE International: Warrendale, PA, USA, 1976.
6. Peris, B.; Navarro-Esbrí, J.; Molés, F. Bottoming organic Rankine cycle configurations to increase Internal Combustion Engines power output from cooling water waste heat recovery. *Appl. Therm. Eng.* **2013**, *61*, 364–371. [CrossRef]
7. Yu, G.; Shu, G.; Tian, H.; Wei, H.; Liu, L. Simulation and thermodynamic analysis of a bottoming Organic Rankine Cycle (ORC) of diesel engine (DE). *Energy* **2013**, *51*, 281–290. [CrossRef]
8. Lu, Y.; Wang, Y.; Dong, C.; Wang, L.; Roskilly, A.P. Design and assessment on a novel integrated system for power and refrigeration using waste heat from diesel engine. *Appl. Therm. Eng.* **2015**, *91*, 591–599. [CrossRef]
9. Vaja, I.; Gambarotta, A. Internal Combustion Engine (ICE) bottoming with Organic Rankine Cycles (ORCs). *Energy* **2010**, *35*, 1084–1093. [CrossRef]
10. Kalina, J. Integrated biomass gasification combined cycle distributed generation plant with reciprocating gas engine and ORC. *Appl. Therm. Eng.* **2011**, *31*, 2829–2840. [CrossRef]

11. Mingshan, W.; Jinli, F.; Chaochen, M.; Noman, D.S. Waste heat recovery from heavy-duty diesel engine exhaust gases by medium temperature ORC system. *Sci. China Technol. Sci.* **2011**, *54*, 2746–2753.

12. Yao, B.; Yang, F.; Zhang, H.; Wang, E.; Yang, K. Analyzing the Performance of a Dual Loop Organic Rankine Cycle System for Waste Heat Recovery of a Heavy-Duty Compressed Natural Gas Engine. *Energies* **2014**, *7*, 7794–7815. [CrossRef]

13. Yang, F.; Zhang, H.; Yu, Z.; Wang, E.; Meng, F.; Liu, H.; Wang, J. Parametric optimization and heat transfer analysis of a dual loop ORC (organic Rankine cycle) system for CNG engine waste heat recovery. *Energy* **2017**, *118*, 753–775. [CrossRef]

14. Yang, F.; Cho, H.; Zhang, H.; Zhang, J. Thermoeconomic multi-objective optimization of a dual loop organic Rankine cycle (ORC) for CNG engine waste heat recovery. *Appl. Energy* **2017**, *205*, 1100–1118. [CrossRef]

15. Wang, E.; Yu, Z.; Zhang, H.; Yang, F. A regenerative supercritical-subcritical dual-loop organic Rankine cycle system for energy recovery from the waste heat of internal combustion engines. *Appl. Energy* **2017**, *190*, 574–590. [CrossRef]

16. Han, Y.; Kang, J.; Wang, X.; Liu, Z.; Tian, J.; Wang, Y. Modelling and simulation analysis of an ORC-FPC waste heat recovery system for the stationary CNG-fuelled compressor. *Appl. Therm. Eng.* **2015**, *87*, 481–490. [CrossRef]

17. Song, S.; Zhang, H.; Zhao, R.; Meng, F.; Liu, H.; Wang, J.; Yao, B. Simulation and Performance Analysis of Organic Rankine Systems for Stationary Compressed Natural Gas Engine. *Energies* **2017**, *10*, 544. [CrossRef]

18. Liu, P.; Shu, G.; Tian, H.; Wang, X. Engine Load Effects on the Energy and Exergy Performance of a Medium Cycle/Organic Rankine Cycle for Exhaust Waste Heat Recovery. *Entropy* **2018**, *20*, 137.

19. Chacartegui, R.; Sánchez, D.; Múñoz, J.; Sánchez, T. Alternative ORC bottoming cycles FOR combined cycle power plants. *Appl. Energy* **2009**, *86*, 2162–2170. [CrossRef]

20. Qiu, K.; Hayden, A. Integrated thermoelectric and organic Rankine cycles for micro-CHP systems. *Appl. Energy* **2012**, *97*, 667–672. [CrossRef]

21. Drescher, U.; Brüggemann, D. Fluid selection for the Organic Rankine Cycle (ORC) in biomass power and heat plants. *Appl. Therm. Eng.* **2007**, *27*, 223–228. [CrossRef]

22. Mago, P.J.; Chamra, L.M.; Srinivasan, K.; Somayaji, C. An examination of regenerative organic Rankine cycles using dry fluids. *Appl. Therm. Eng.* **2008**, *28*, 998–1007. [CrossRef]

23. Kosmadakis, G.; Manolakos, D.; Kyritsis, S.; Papadakis, G. Comparative thermodynamic study of refrigerants to select the best for use in the high-temperature stage of a two-stage organic Rankine cycle for RO desalination. *Desalination* **2009**, *243*, 74–94. [CrossRef]

24. Tchanche, B.F.; Papadakis, G.; Lambrinos, G.; Frangoudakis, A. Fluid selection for a low-temperature solar organic Rankine cycle. *Appl. Therm. Eng.* **2009**, *29*, 2468–2476. [CrossRef]

25. Tian, H.; Shu, G.; Wei, H.; Liang, X.; Liu, L. Fluids and parameters optimization for the organic Rankine cycles (ORCs) used in exhaust heat recovery of Internal Combustion Engine (ICE). *Energy* **2012**, *47*, 125–136. [CrossRef]

26. Hung, T.-C.; Lee, D.-S.; Lin, J.-R. An Innovative Application of a Solar Storage Wall Combined with the Low-Temperature Organic Rankine Cycle. *Int. J. Photoenergy* **2014**, *2014*, 1–12. [CrossRef]

27. Zare, V. A comparative exergoeconomic analysis of different ORC configurations for binary geothermal power plants. *Energy Convers. Manag.* **2015**, *105*, 127–138. [CrossRef]

28. Fontalvo, A.; Solano, J.; Pedraza, C.; Bula, A.; Quiroga, A.G.; Padilla, R.V. Energy, Exergy and Economic Evaluation Comparison of Small-Scale Single and Dual Pressure Organic Rankine Cycles Integrated with Low-Grade Heat Sources. *Entropy* **2017**, *19*, 476. [CrossRef]

29. Calise, F.; D'Accadia, M.D.; Macaluso, A.; Piacentino, A.; Vanoli, L. Exergetic and exergoeconomic analysis of a novel hybrid solar–geothermal polygeneration system producing energy and water. *Energy Convers. Manag.* **2016**, *115*, 200–220. [CrossRef]

30. Kerme, E.D.; Orfi, J. Exergy-based thermodynamic analysis of solar driven organic Rankine cycle. *J. Therm. Eng.* **2015**, *1*, 192–202. [CrossRef]

31. Jouhara, H.; Sayegh, M.A. Energy efficient thermal systems and processes. *Therm. Sci. Eng. Prog.* **2018**, *7*, e1–e2. [CrossRef]

32. Chen, T.; Zhuge, W.; Zhang, Y.; Zhang, L. A novel cascade organic Rankine cycle (ORC) system for waste heat recovery of truck diesel engines. *Energy Convers. Manag.* **2017**, *138*, 210–223. [CrossRef]

33. Moran, M.J.; Saphiro, H.N.; Boettner, D.D.; Bailey, M.B. *Fundamentals of Engineering Thermodynamics*; Wiley. New York, NY, USA, 2011.
34. Quoilin, S.; Aumann, R.; Grill, A.; Schuster, A.; Lemort, V.; Spliethoff, H. Dynamic modeling and optimal control strategy of waste heat recovery Organic Rankine Cycles. *Appl. Energy* **2011**, *88*, 2183–2190. [CrossRef]
35. Mathworks. *Matlab: Computer Program*; The MatlabWorks Inc.: Denver, CO, USA, 2018.
36. Lemmon, E.W.; Huber, M.L.; McLinden, M.O. *NIST Standard Reference Database 23: Reference Fluid Thermodynamic and Transport Properties-REFPROP, Version 9.1.*; Technical Report; National Institute of Standards and Technology: Gaithersburg, MD, USA, 2013.
37. Trindade, A.B.; Palacio, J.C.E.; González, A.M.; Orozco, D.J.R.; Lora, E.E.S.; Renó, M.L.G.; Del Olmo, O.A. Advanced exergy analysis and environmental assesment of the steam cycle of an incineration system of municipal solid waste with energy recovery. *Energy Convers. Manag.* **2018**, *157*, 195–214. [CrossRef]
38. Val, C.G.F.; De Oliveira, S., Jr. Deep Water Cooled ORC for Offshore Floating Oil Platform Applications. *Int. J. Thermodyn.* **2017**, *20*, 229–237. [CrossRef]
39. El-Emam, R.S.; Dincer, I. Exergy and exergoeconomic analyses and optimization of geothermal organic Rankine cycle. *Appl. Therm. Eng.* **2013**, *59*, 435–444. [CrossRef]
40. Uusitalo, A.; Honkatukia, J.; Turunen-Saaresti, T.; Larjola, J. A thermodynamic analysis of waste heat recovery from reciprocating engine power plants by means of Organic Rankine Cycles. *Appl. Therm. Eng.* **2014**, *70*, 33–41. [CrossRef]
41. Feng, Y.; Zhang, Y.; Li, B.; Yang, J.; Shi, Y. Comparison between regenerative organic Rankine cycle (RORC) and basic organic Rankine cycle (BORC) based on thermoeconomic multi-objective optimization considering exergy efficiency and levelized energy cost (LEC). *Energy Convers. Manag.* **2015**, *96*, 58–71. [CrossRef]
42. Lai, N.A.; Wendland, M.; Fischer, J. Working fluids for high-temperature organic Rankine cycles. *Energy* **2011**, *36*, 199–211. [CrossRef]
43. Desai, N.B.; Bandyopadhyay, S. Process integration of organic Rankine cycle. *Energy* **2009**, *34*, 1674–1686. [CrossRef]
44. Rayegan, R.; Tao, Y.X. A procedure to select working fl uids for Solar Organic Rankine Cycles (ORCs). *Renew. Energy* **2011**, *36*, 659–670. [CrossRef]
45. Guzovi, Z. The comparision of a basic and a dual-pressure ORC (Organic Rankine Cycle): Geothermal Power Plant Velika Ciglena case study. *Energy* **2015**, *76*, 175–186. [CrossRef]
46. Minea, V. Power generation with ORC machines using low-grade waste heat or renewable energy. *Appl. Therm. Eng.* **2014**, *69*, 143–154. [CrossRef]
47. Nami, H.; Mohammadkhani, F.; Ranjbar, F. Utilization of waste heat from GTMHR for hydrogen generation via combination of organic Rankine cycles and PEM electrolysis. *Energy Convers. Manag.* **2016**, *127*, 589–598. [CrossRef]

Article

Optimum Organic Rankine Cycle Design for the Application in a CHP Unit Feeding a District Heating Network

Lisa Branchini, Andrea De Pascale *, Francesco Melino and Noemi Torricelli

Department of Industrial Engineering, Alma Mater Studiorum—Università di Bologna, viale del Risorgimento 2, 40136 Bologna, Italy; lisa.branchini2@unibo.it (L.B.); francesco.melino@unibo.it (F.M.); noemi.torricelli2@unibo.it (N.T.)
* Correspondence: andrea.depascale@unibo.it; Tel.: +39-051-2093310

Received: 28 February 2020; Accepted: 10 March 2020; Published: 12 March 2020

Abstract: Improvement of energy conversion efficiency in prime movers has become of fundamental importance in order to respect EU 2020 targets. In this context, hybrid power plants comprising combined heat and power (CHP) prime movers integrated with the organic Rankine cycle (ORC) create interesting opportunities to additionally increase the first law efficiency and flexibility of the system. The possibility of adding supplementary electric energy production to a CHP system, by converting the prime movers' exhaust heat with an ORC, was investigated. The inclusion of the ORC allowed operating the prime movers at full-load (thus at their maximum efficiency), regardless of the heat demand, without dissipating not required high enthalpy-heat. Indeed, discharged heat was recovered by the ORC to produce additional electric power at high efficiency. The CHP plant in its original arrangement (comprising three internal combustion engines of 8.5 MW size each) was compared to a new one, involving an ORC, assuming three different layout configurations and thus different ORC off-design working conditions at user thermal part-load operation. Results showed that the performance of the ORC, on the year basis, strongly depended on its part-load behavior and on its regulation limits. Indeed, the layout that allowed to produce the maximum amount of ORC electric energy per year (about 10 GWh/year) was the one that could operate for the greatest number of hours during the year, which was different from the one that exhibited the highest ORC design power. However, energetic analysis demonstrated that all the proposed solutions granted to reduce the global primary energy consumption of about 18%, and they all proved to be a good investment since they allowed to return on the investment in barely 5 years, by selling the electric energy at a minimum price equal to 70 EUR/MWh.

Keywords: organic Rankine cycle; waste heat recovery; internal combustion engine; cogeneration; district heating

---

1. Introduction

In the framework of the "20–20–20" targets [1], measures adopted by the European Union to mitigate climate change include the improvement of conventional prime movers' conversion efficiency. As indicated in the European directive 2012/27/EU [2], the promotion of cogeneration, as well as waste heat recovery (WHR) strategies, could significantly contribute to achieving this objective.

Internal combustion engines typically convert just the 30–45% of the fuel primary energy into mechanical or electric power, while the remaining part is rejected as thermal energy with exhaust gases, cooling water, oil, etc. This heat, which would be wasted otherwise, can be used for combined heat and power (CHP) applications, thus improving the overall first law conversion efficiency, but also for additional electric production, via a bottomer thermodynamic cycle. Heat released by internal combustion engine (ICE) exhaust gases is usually characterized by (i) limited thermal power size (varying typically in the range from kW to tens of MW as order of magnitude), (ii) by temperature often quite lower than 500 °C, and (iii) both strongly variable with the engine load. These conditions are not always compatible with the adoption of conventional steam Rankine cycle [3] since problems related to the use of water as working fluid can arise as (i) difficulty in superheating the fluid in order to avoid condensate formation during the expansion process and the risk of erosion of the turbine blades [3]; (ii) working with excessive vapor operating pressure that impose to install complex multi-stage and expensive turbines [3]. A solution is identified in changing the operating fluid with an organic compound in an organic Rankine cycle (ORC). Organic fluids are characterized by higher molecular mass and lower critical temperature compared to water that allows exploiting efficiently, also, low-grade heat sources. Even though the conversion efficiency of ORC is typically limited, compared to steam cycle, this technology features a number of advantages: lower O&M and personnel costs; high molecular weight; low enthalpy drop in the expander, and, as a consequence, higher mass flow rates, if compared with water. Moreover, ORCs are characterized by a wider regulation range, and performances are not particularly penalized at part-load conditions [4].

WHR from stationary ICE prime movers is quite a consolidated option, and several investigations have been proposed on this topic (see for example [5–14]). The studies point out that ORC performance is significantly affected by the choice of the working fluid, the cycle architecture, and the setting of the operating parameters. Investigated fluids are hydrocarbons, refrigerants, and zeotropic mixtures, while the other most important decision variables are identified in the evaporating pressure and the superheating degree [5–7]. In particular, Yang et al. analyzed an ORC system in [5] to recover waste heat from the diesel engine exhaust, where the zeotropic mixture R416A was used as the working fluid for the ORC. Considering various operating conditions of the diesel engine, this study investigated the effects of the degree of superheating on the running performance of the ORC waste heat recovery. In the thermodynamic study by Han et al. [6], a regenerative ORC system was established to recover the waste flue gas of 160 °C, focusing on thermodynamic and economic performance while simultaneously considering the effect of superheating and working fluid selection. The optimization of the evaporation temperature was carried out by analyzing the variation of net power output and specific investment cost. Energy and exergy analysis of three ORC–WHR configurations that use a coupling thermal oil circuit was performed by Valencia et al. in [7], where a simple ORC, an ORC with a recuperator, and an ORC with double-pressure configuration were considered; cyclohexane, toluene, and acetone were simulated as ORC working fluids. In addition, the effect of evaporating pressure on the net power output, thermal efficiency increase, specific fuel consumption, overall energy conversion efficiency, and exergy destruction was also investigated.

These studies demonstrated also the importance of taking into account the part-load operation of the ICEs when evaluating the performance of the ORC bottoming solution. For example, in [8] and [9], the performance of combined systems with ORC and ICE was analyzed at steady-state under different typical working conditions, by means of a numerical simulation model of the system. These papers analyzed, in addition, the effect of adjustable parameters on the system performance, giving effective control directions under various conditions. They proved that to get a better system performance under different working conditions, the system should be operated with a slight degree of superheating.

Valorization of waste heat through ORC creates interesting opportunities, especially to combined heat and power generation, in the residential sector and in energy-intensive industries. In this context, a series of hybrid power plants comprising a CHP prime mover integrated with ORC have been investigated [10–13], with the aim of maximizing the share of electricity production of CHP plants rather than a higher heat production rate in a cost-effective manner. The analyzed heat sources are usually cooling water from steam cycles or exhaust gas from the industrial process. An example of a hybrid CHP plant was proposed by Yi et al. in [10], where an ORC and a hydraulic turbine were introduced to reduce exergy losses, which occurred in a thermal storage process for district heating (DH) purpose.

Results showed a payback period equal to just 3.5 years. The study proposed by Arabkoohsara and Namib [11] analyzed a waste-driven CHP plant. In addition to the electricity generated by the main steam cycle, the heat withdrawn from the condenser was employed for additional heat and electricity production by employing a small-scale ORC unit and a heat exchanger connected to the local district heating system. This study evaluated an electric efficiency improvement of 20% and a favorable period of return on investment. Marty et al. [12] analyzed the optimization of parallel distribution between electricity and heat production for a geothermal plant. The geothermal fluid was split into two streams, one used for an ORC system, and the other for the DH. An optimization tool was implemented to obtain the sizing of the ORC and the best distribution between electricity and heat production. In the project by Ramireza et al. [13], a large-scale ORC pilot plant along with a waste heat recovery unit in a steel mill was designed. Waste heat was recovered from the fumes of the furnace to produce saturated steam, which was then delivered to a DH network during the cold season and to the ORC for electricity generation during the rest of the year. Results on the performance of the plant during the first months of operation showed promising values. The study by Grljušić et al. [14] showed instead an application involving a diesel engine for maritime application. A standard operating propulsion engine at maximum efficiency was assumed to supply the auxiliary power. The aim of this research was to investigate the possibility of using the CHP plant over a ship to meet all heating and electricity requirements during navigation.

All these studies make evident that the ORC design for CHP hybrid systems has to be determined in accordance with the specific application. In particular, in the case of a CHP ICE-ORC hybrid system, the optimal ORC sizing, the optimal load allocation, and the resulting off-design performance should be explored more in depth, especially if it is called to operate under variable thermal demand conditions.

The Objective of the Study

The main objective of this work was the energy and economic assessment of a CHP power plant based on ICEs, feeding a DH network, and coupled with an ORC to optimally recover the ICEs' residual heat. In this work, a specific case study of an existing CHP system was analyzed; however, the same presented approach and similar considerations could be applied to any CHP feeding a district heating network (DHN) as a thermal user.

In detail, as the first step, the optimal operation of the CHP power plant was assessed without ORC: the ICEs load profile was determined, in order to fulfill the DH thermal demand, while minimizing the heat dissipation to the ambient. Then, in order to increase the amount of produced electrical energy and to better exploit the available ICEs residual heat, the opportunity of integrating an ORC as a bottomer was investigated.

The original power plant optimal operation was compared with the new arrangement, where the ICEs were operated at full-load, and the ORC was integrated also with the DH demand, to recover the ICEs' residual heat. The ORC as bottomer of the CHP was an interesting solution also because it could represent a flexibilization tool in the regulation of the whole plant. The ORC allowed operating the ICEs at full-load (i.e., at maximum efficiency point) while not dissipating the surplus of heat production when it is not required by the thermal user. Indeed, the surplus of high enthalpy-heat production was recovered by the ORC and converted into additional electric power, sold to the electric grid.

In detail, three different system architectures were proposed and analyzed. A sensitivity analysis of organic fluid and key cycle parameters was performed in order to identify, for each proposed arrangement, the optimum design of the ORC. A thermodynamic design and off-design model of the ORC system has been developed in the *Thermoflex*™ environment [15]. The additional electrical energy produced during the year, the ORC operating hours, and total investment costs were quantified and compared among the analyzed layouts. Finally, the original optimal operation of the CHP plant was compared with the modified one comprising the ORC, in order to estimate, for each proposed arrangement, the amount of additional generated electric power and the corresponding increase in primary energy consumption.

An energetic index accounting for the global primary energy saving was introduced, in order to evaluate the convenience in producing the additional electric power with the modified CHP plant rather than with the original plant with the help of the electric generating mix. From the economic point of view, considerations on the electric energy sell price to return on investment were also presented.

2. Cogeneration Power Plant Description

This section presents the description of the CHP plant feeding the DH network and the control logic adapted to regulate the system when working at part-load conditions.

2.1. CHP Plant Arrangement

The case study system was an existing cogeneration plant intended to satisfy the heat demand of an international hub airport; it comprised three ICEs (model Rolls Royce Bergen engine B35: 40 V20 AG), fed with natural gas, named GR1, GR2, and GR3, as depicted in Figure 1. The CHP single unit prime movers were conceived to provide a nominal electric power equal to 8.5 MW and thermal power of about 4.6 MW (only heat recovery from the exhaust gas is considered). Thus, the nominal electric and thermal efficiency were equal to 47% and 25%, respectively. The DH demand was fulfilled by the ICEs discharged heat and by three additional boilers, producing 8 MW of nominal thermal power output each, also fed with natural gas. The generated electric power was fed into the grid at a connection voltage of 150 kV.

The DH network (DHN) was designed to provide hot water for various types of tertiary building users. It included heat exchangers for heating purposes in winter and heat exchangers working all along the year, providing high-temperature heat to produce food use steam and to feed absorption refrigeration units, but also sanitary water.

For what regards the management strategy, a DHN could be regulated (i) varying the mass flow rate but keeping constant the temperature difference (ΔT) on the utilities (this approach is particularly suitable for small size DHNs) or (ii) varying the ΔT on the utilities with a constant mass flow rate (usually for large DHNs) [16]. The regulation adopted in this study was the first one: in particular, it was assumed that the delivery temperature was kept equal to 130 °C and the return equal to 70 °C. This supposition could be reasonable if considering that the DHN inertia was typically such high to presume to be neglectable the return temperature oscillation when relying on an hourly basis (which was the timestep considered in this analysis).

The yearly required thermal production profile is shown in Figure 2. Higher thermal power requests were observed during winter, with a demand peak close to 20 MW; lower heat demand values, around 3 MW on average, were registered during summer months.

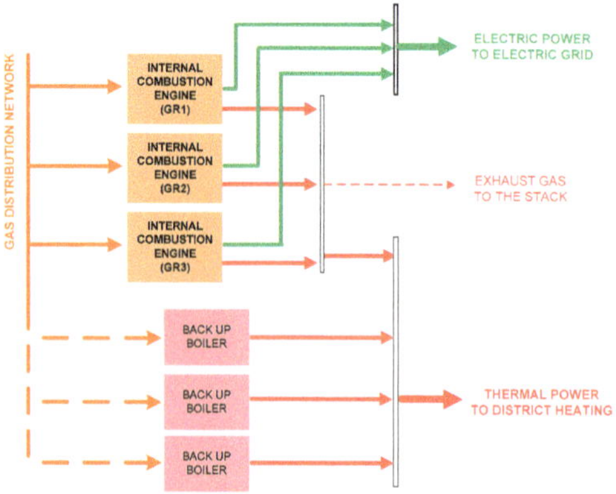

Figure 1. Schematic of the cogeneration power plant.

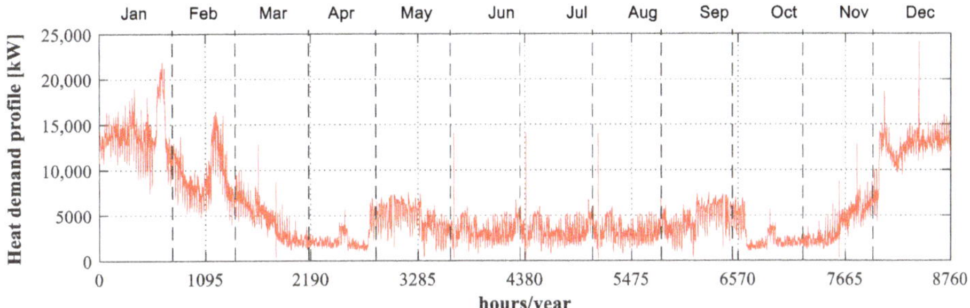

Figure 2. Heat demand yearly profile.

2.2. CHP Plant Control Logic

The control logic of the CHP plant was defined based on the result of an optimal load allocation procedure, which led to defining the optimal ICE single unit load for each value of the heat demand. The optimal load allocation problem was formulated as follows (Equation (1)): CHP prime movers were managed in order to cover the DH demand ($\dot{Q}_{th,DH}$) while keeping equal to zero the heat dissipation to stack ($\dot{Q}_{th,Diss}$), with the following constraints: (i) the ICE regulation ranges in between 40% to 100% of the nominal load (according to ICE datasheet); (ii) minimum exhaust gas temperature ($T_{exhaust}$) equal to 110 °C [17], as conservative value, to prevent water condensation along the discharge line; (iii) when a potential ICE heat dissipation would occur, the boilers are turned on to cover the heat demand, in place of the ICEs, in order to avoid high-grade enthalpy wasted heat.

$$\dot{Q}_{th,Diss}(\overline{x}) = 0 \; such \; that \begin{cases} \dot{Q}_{th,PM}(\overline{x}) = \dot{Q}_{th,DH} \; if \; \dot{Q}_{th,DH} < \dot{Q}_{th,PM,max} \\ \dot{Q}_{th,PM}(\overline{x}) = \dot{Q}_{th,PM,max} \; if \; \dot{Q}_{th,DH} \geq \dot{Q}_{th,PM,max} \\ T_{exhaust}(\overline{x}) = 110 \, °C \\ 0.4 \; \leq \overline{x} \leq 1 \end{cases} \tag{1}$$

$$\overline{x} = load_{GR1}, load_{GR2}, load_{GR3}$$

where the problem control variables ($\overline{x}$) are the three ICE loads ($load_{GR1}$, $load_{GR2}$, $load_{GR3}$). The resulting optimization problem could be treated as a constrained, multivariable function, minimization problem.

The problem was solved considering the heat demand profile (Figure 2) and the ICE and boiler performance maps (Figure 3). The boiler efficiency trend describes the part-load behavior of a boiler conceived for the auxiliary purpose [18], thus characterized by high performance, even at very low loads. In particular, Figure 3 shows the ICE exhaust gas mass flow rate and temperature, the electrical and thermal power output versus load (data provided by the manufacturer). The thermal power was calculated, assuming cool exhaust gases down to 110 °C. It resulted that each ICE could be regulated between the 40 and the 100% of its load, covering a heat demand ranging between 3200 and 4600 kW; thus, the three ICEs together could cover a maximum heat demand equal to 13,800 kW.

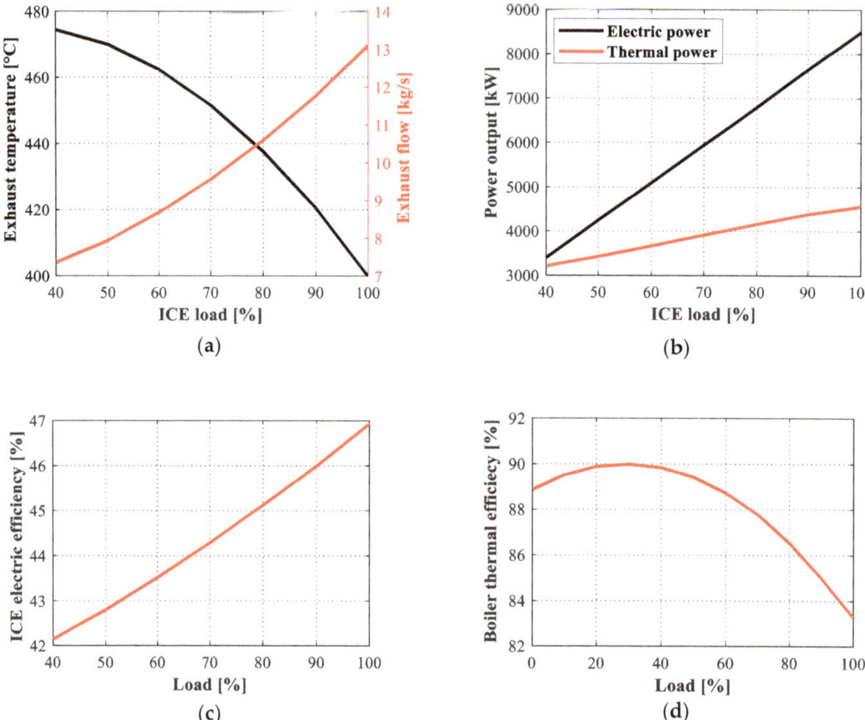

Figure 3. Performance of internal combustion engines (ICEs) and back-up boilers at part-load conditions: (**a**) ICE exhaust gas conditions at part-load operation; (**b**) ICE electric and thermal power output at part-load operation; (**c**) ICE electric efficiency at part-load operation; (**d**) Boilers thermal efficiency at part-load operation.

Results of optimal load allocation procedure are reported in Figure 4, where the load of the three ICE units (Figure 4a), the CHP unit total electric power output (Figure 4b), and boilers load (Figure 4c) as a function of the heat demand are shown. Finally, in Figure 4d, the total natural gas consumption (due to ICE units and boilers) was plotted versus the heat demand. It could be observed that, in the range of power where the ICEs operated, only the least number of units necessary to cover the demand was activated. They were turned on at their minimum load, and then the load increased with the heat demand until reaching 100%. Beyond 13,800 kW, all the ICEs worked at full-load, but the generated power was not enough to cover the heat demand; therefore, also the back-up boilers were activated. The boilers were activated also at heat demand lower than 13,800 kW when the ICEs could not fulfill the heat demand due to the regulation limits expressed in (1).

The total amount of natural gas consumed to fulfill the DH yearly heat demand resulted equal to 138.36 GWh/year; it guaranteed to cover the heat demand and to deliver 54.22 GWh/year of electric power to the grid. The ICE and the boiler fuel consumption were evaluated on the basis of the efficiency maps reported in Figure 3c,d.

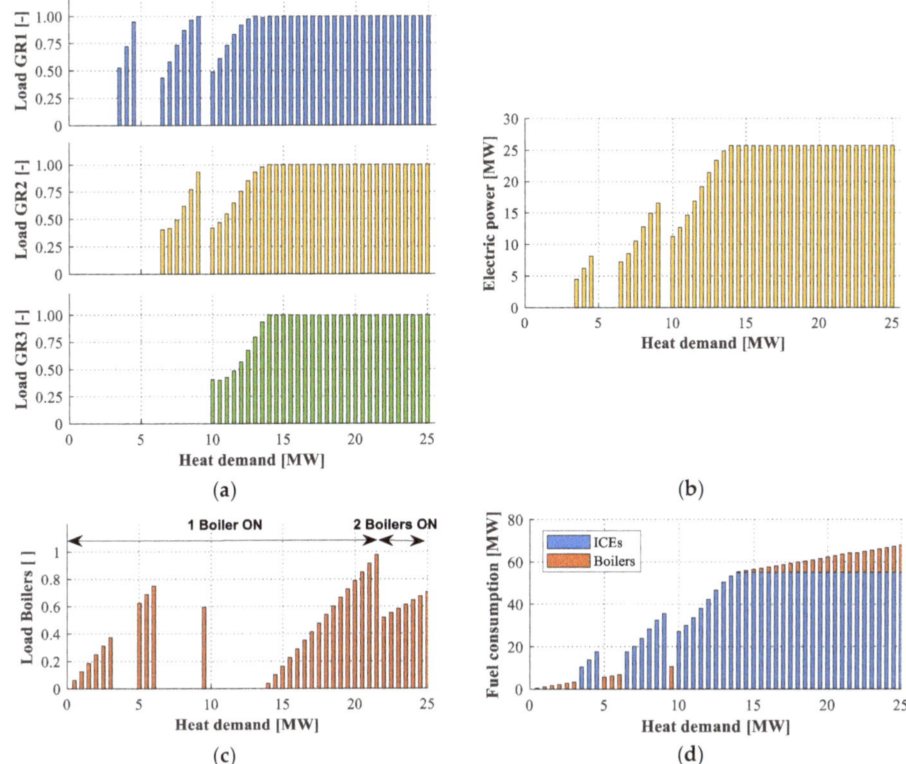

Figure 4. Combined heat and power (CHP) system output loads and fuel consumption as a function of the district heating (DH) heat demand: (**a**) load of the ICE units; (**b**) CHP unit total electric power output; (**c**) boilers load as a function of the heat demand; (**d**) total natural gas consumption.

3. Integrating the Organic Rankine Cycle

In order to increase the amount of electrical energy sold to the grid, the opportunity of integrating an ORC as a bottoming cycle of the ICEs was considered. Since ORC technology offers the possibility to generate electricity from low-grade heat sources, the residual heat of ICEs exhausts flue gases was here considered as a potential feeding source. Thus, the CHP plant optimal operation was compared with a new arrangement where ICEs were operated at full-load, and an ORC was integrated into the layout to recover ICEs' residual heat. Three different ICE-ORC layouts were proposed and analyzed. For each investigated arrangement, optimum ORC design (i.e., selection of organic fluid and key cycle parameters) was identified; based on ORC system off-design modeling, the amount of additional electrical energy producible during the year was then quantified.

3.1. ORC Architecture

A simple sub-critical and regenerative ORC architecture with an intermediate heat transfer fluid (IHTF) loop was assumed, as schematically shown in Figure 5. The key cycle components were: (i) the heat exchanger between ICE exhaust gases and heat carrier fluid, (ii) the evaporator (exchanging heat between the heat carrier fluid and organic fluid), (iii) the expander, (iv) the regenerative heat exchanger, and (v) the condenser and pumps. The use of an intermediate thermal circuit, which could act as buffer, was crucial for the application under consideration in order to: (i) avoid the direct contact between exhaust gases and flammable organic fluid (such as hydrocarbon), (ii) smooth out the variation of temperature and mass flow rate of the heat source.

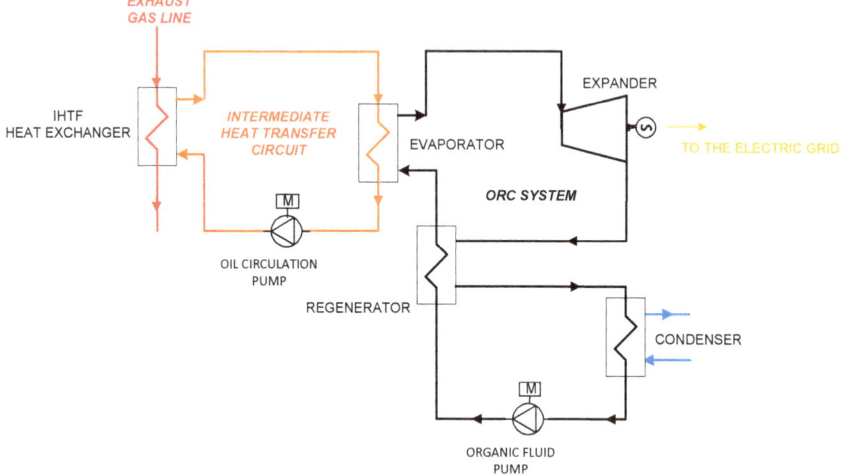

Figure 5. Schematic of the organic Rankine cycle (ORC) architecture.

3.2. Proposed Series and Parallel Arrangements

Three different layouts were investigated in this work, as schematically shown in Figure 6.

In the first case (Case A), the ORC was placed downstream the heat exchanger feeding the DH network, thus recovering the residual heat of exhausted gases. Therefore, the DH heat demand would determine the temperature of the gas feeding the ORC.

In the second architecture (Case B), the heat recovery heat exchanger position was reversed, with the ORC located upstream of the DH heat exchanger. Accordingly, in this second arrangement, the ORC would be fed by a constant temperature, equal to 400 °C (i.e., the temperature of ICE exhausted at full-load condition, see Figure 3a), while the total amount of heat input to the cycle would vary, according to the DH request.

In the third analyzed layout (Case C), a parallel arrangement was assumed: exhaust gas flow would be split between the two branches in order to primarily satisfy the DH load demand, while both branches featured the same inlet gas temperature value (equal to 400 °C). A collector was placed downstream components, in order to mix up the two different fluxes; the minimum requirement of 110 °C on temperature was maintained in section G3 of Figure 6.

Indeed, for all the above described integrated system architectures, the heat demand of DH was the primary requirement to meet. Consequently, the thermal power input to the ORC equaled the difference between the available thermal power in the ICE exhaust gas and heat delivered to the DH network ($\dot{Q}_{th,DH}$). For each investigated layout, Table 1 collects the assumed constant and variable parameters (see the section in Figure 6). The intermediate heat transfer fluid heat exchanger, which provided the thermal power input to the ORC, was indicated as IHTF HE for the sake of brevity.

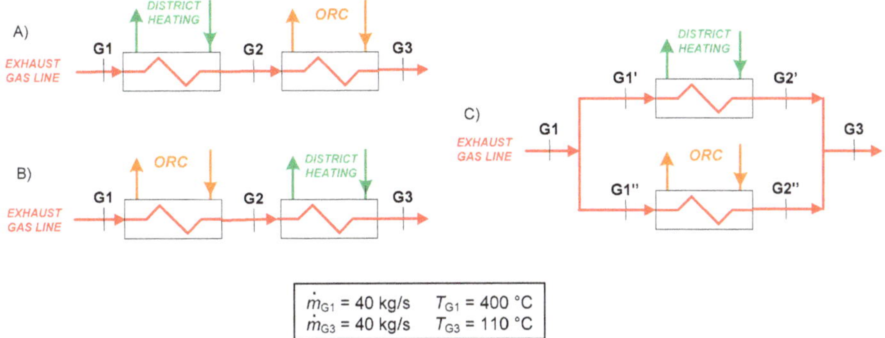

Figure 6. Schematic of the proposed DH and intermediate heat transfer fluid (IHTF) heat exchangers arrangements: (**a**) case A; (**b**) case B; (**c**) case C.

Table 1. Constants and variable parameters for each proposed layout.

Layouts	Case A	Case B	Case C
Exhaust gas mass flow rate entering the IHTF HE	$\dot{m}_{G2} = 40\,\text{kg/s}$	$\dot{m}_{G1} = 40\,\text{kg/s}$	$\dot{m}_{G1''} = f\left(\dot{Q}_{th,DH}\right)$
Exhaust gas temperature entering the IHTF HE	$T_{G2} = f\left(\dot{Q}_{th,DH}\right)$	$T_{G1} = 400\,°\text{C}$	$T_{G1''} = 400\,°\text{C}$
Outlet gas temperature at the IHTF HE	$T_{G3} = 110\,°\text{C}$	$T_{G2} = f\left(\dot{Q}_{th,DH}\right)$	$T_{G2''} = 110\,°\text{C}$

3.3. Selected Organic Working Fluids

Four different hydrocarbons (Cyclopentane, Benzene, Cyclohexane, and Toluene) were selected as potential working fluids, as demonstrated to be performing fluids for recovering heat at the temperature imposed by the ICEs exhausts [19,20]. Indeed, the selected fluids exhibited a critical temperature quite similar or slightly higher than the target evaporation temperature, in particular high enough to achieve a good thermal matching between fluids and exhaust gas, but not too high to lead to excessively low vapor densities and thus high system cost [21]. Refrigerants were excluded due to their low critical temperature, while siloxanes were excluded because their use would lead to excessively low saturation pressure values, assuming cooling fluid temperature close to ambient conditions [21]. The main properties of selected fluids are summarized in Tables 2 and 3, and the fluids saturation pressure values are plotted versus temperature for comparative purpose in Figure 7.

The IHTF considered was Therminol 62, one of the most popular high-temperature liquid phase heat transfer fluid. Therminol 62 was selected as it offers outstanding performance to 325 °C, including excellent thermal stability and low vapor pressure. These properties result in reliable, consistent performance of heat transfer systems over long periods of time (main properties are reported in Table 3) [22].

Table 2. Main properties of selected organic fluids [21].

Organic Fluid Property	Cyclopentane	Benzene	Cyclohexane	Toluene
Molar weight (kg/kmol)	70.1	78.1	84.2	92.1
Critical pressure (bar)	45.8	49.1	40.8	41.3
Critical temperature (°C)	238.6	288.9	280.5	318.6
Normal boiling point (°C)	49.3	80.1	80.7	110.6
Autoignition temperature (°C)	320	555	260	480

Table 3. Main properties of selected IHTF [22].

IHTF Property	Therminol 62
Molar weight (kg/kmol)	252
Normal boiling point (°C)	333
Autoignition temperature (°C)	433
Liquid density at 25 °C (kg/m^3)	951

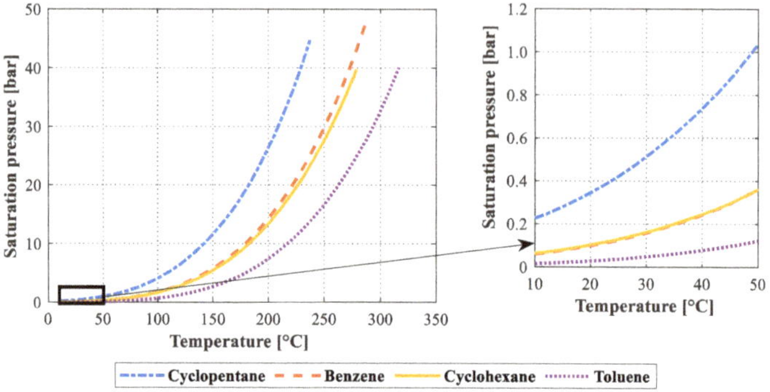

Figure 7. Pressure-temperature saturation curves of selected hydrocarbons.

3.4. ORC Thermodynamic Modeling and Design Assumptions

The ORC system and the IHTF circuit were modeled and simulated by means of the commercial software *Thermoflex™* [15]. The software, integrated with the FluidProp library for organic fluids thermodynamic properties evaluation, allowed simulating the energy system steady state, based on a lumped parameters approach.

First of all, the whole system layout was reproduced in the *Thermoflex™* environment, combining the built-in library single component modules of heat exchangers, turbines, pumps, sources, and sinks. Boundary conditions and components design key parameter values used in the study (such as turbine isentropic efficiency, condensation pressure, pressure drops, and heat losses along the circuit) are listed in Tables 4–6, mainly selected in line with the state-of-the-art of conventional ORC technology [23].

Table 4. ORC layouts dependent assumptions.

Layout CASE	A1	A2	A3	B	C
Exhaust gas temperature entering the IHTF HE (°C)	342 (G2)	342 (G2)	342 (G2)	400 (G1)	400 (G1″)
Exhaust gas mass flow rate entering the IHTF HE (kg/s)	40 (G2)	40 (G2)	40 (G2)	40 (G1)	32 (G1″)
Outlet gas temperature exiting the IHTF HE (°C)	110 (G3)	110 (G3)	110 (G3)	208 (G2)	110 (G2″)
IHTF temperature (T_{des}) (°C)	300	250	200	300	300

Table 5. ORC general assumptions.

General Assumption	
Pressure drop at the heat exchangers (%)	10
Heat exchangers thermal loss (%)	1
IHTF loop max. pressure (bar)	15 15
Pressure drop in IHTF loop (%)	15
ORC expander isentropic efficiency (%)	80 80
Regenerator effectiveness (%)	85
Subcooling ORC outlet condenser (°C)	5
Pumps nominal isentropic efficiency (%)	60

Table 6. ORC condensation pressure assumptions.

Fluids	Condensation Pressure (Bar)
Cyclopentane	0.74
Benzene	0.24
Toluene	0.08
Cyclohexane	0.25

In order to perform a realistic evaluation of the ORC performance, both thermodynamic design and off-design analyses were performed. Off-design performance of the plants were obtained passing through three main steps: (i) preliminary thermodynamic heat and mass balances; (ii) engineering design phase, in which design parameters of the power plant are defined (i.e., size of components, geometric details, etc.); (iii) thus, off-design analysis where, being defined the design characteristics and fixed the geometry of the components, depending on selected control logic, it is possible to predict both components and overall system behavior under different operating conditions.

The main equations used to calculate streams' design parameters included energy balances at the single components. A more exhaustive description of the Thermoflex modeling approach can be found in [23]. Concerning the off-design modeling approach, the heat exchangers off-design behavior was calculated according to the method of thermal resistance scaling [15]. Being determined as the design-point overall thermal resistance of the heat exchanger, the two fluid-side resistances were scaled at off-design, using the single-parameter scaling method [15]. Normalized heat loss in each heat exchanger was considered, and it was entered as a percentage, relative to the heat transferred out of the higher temperature fluid. Flow resistance coefficients, initialized in the design-point stage, were used to model the pressure drops across each side of the heat exchanger in off-design. Regarding the ORC expander, the sliding pressure part-load control was assumed to model its behavior at off-design conditions. Thus, whenever possible (based on topper off-design performance), the organic fluid temperature at the expander inlet was kept constant at its rated design value, while mass flow and pressure varied proportionally assuming chocking conditions (constant mass flow function) at the turbine inlet. Expander isentropic efficiency at off-design was corrected, starting from the design point value, based on flow function.

Table 4 shows the selected design values for the inlet/outlet temperature and mass flow of the gas at the IHTF HE, for the different proposed layouts; these values corresponded to a DH demand equal to 3 MW, the mean value of the thermal demand in the period of minimum request (May to September). The IHTF circuit design temperature at the HE outlet (T_{des}) was set equal to 300 °C in Case B and Case C layouts, while three possible values (Case A1: 200 °C, Case A2: 250 °C, and Case A3: 300 °C) were considered in Case A layout, where the IHTF HE was downstream the DH HE.

The condensation pressure depends on the working fluid, and it is bound by the cold source temperature. In this study, the cooling medium temperature was assumed equal to 25 °C as representative of a medium climate condition [24]. Considering a reasonable temperature difference between organic fluid saturation temperature and cooling medium temperature inlet equal to 15 °C, the organic fluid condensation temperature was set equal to 40 °C for all considered fluids, resulting in the condensation pressure values in Table 6 (see also Figure 7). In order to guarantee a complete fluid condensation inside the condenser, a subcooling equal to 5 °C was assumed.

4. Results and Discussion

4.1. Optimization of the Cycle Parameter and Working Fluid Selection

A preliminary thermodynamic sensitivity analysis was performed on the considered working fluids, in order to optimize the cycle evaporation pressure and then determine the fluid that maximizes the net electric power production, for each analyzed arrangement.

The influence of organic fluid and key cycle operating parameters are presented in Figure 8, with reference to Case A layout. Since the exhaust gas entering the IHTF HE was variable in case A (see Table 1), a comparative analysis of the net power output is presented in Figure 8 for each analyzed T_{des} value and for each organic fluid. Results showed that Cyclopentane achieved the best performance compared to other hydrocarbons, for all the considered T_{des} values. However, it must be pointed out that ORC performance was reduced by decreasing the IHTF temperature. Optimum evaporation pressure values could be identified equal to 40, 16, and 10 bar, respectively, at 300, 250, and 200 °C. As expected, the highest ORC performance was achieved for the highest temperature (i.e., 1490 kW of net power output for Cyclopentane with a pressure equal to 40 bar and T_{des} equal to 300 °C). If the design IHTF temperature was reduced to 200 °C, the maximum ORC power output was reduced by nearly 20%.

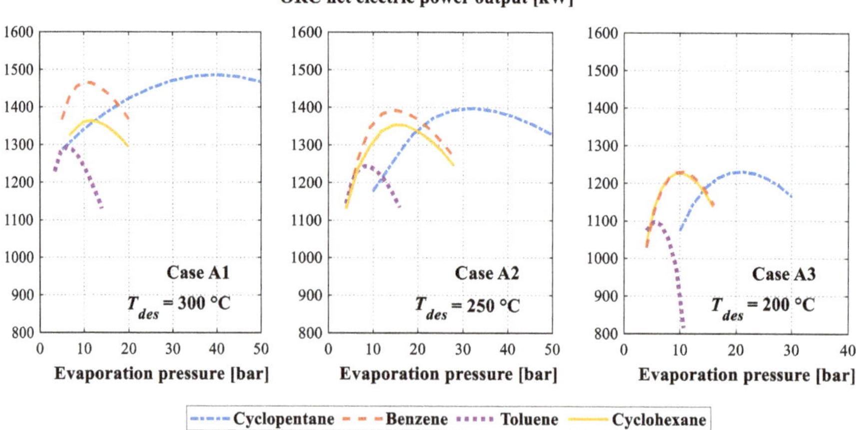

Figure 8. Sensitivity analysis results on organic fluid and IHTF design temperature in Case A layout and different IHFT temperature cases.

Figure 9 shows the results for Case B and Case C: a fixed IHTF design temperature value equal to 300 °C was considered. Indeed, these proposed layouts featured the maximum temperature of exhaust gas entering the IHTF HE (T_{G1} and T_{G1}'', respectively, see Table 1). Cyclopentane still provided the best performance in Case B and Case C layouts. The maximum power output was achieved with evaporation pressure equal to 40 bar for both layout arrangements. The highest ORC power output value was obtained in Case B, in which the IHTF HE was fed with the exhaust gas at its highest temperature and mass flow if compared with the other cases.

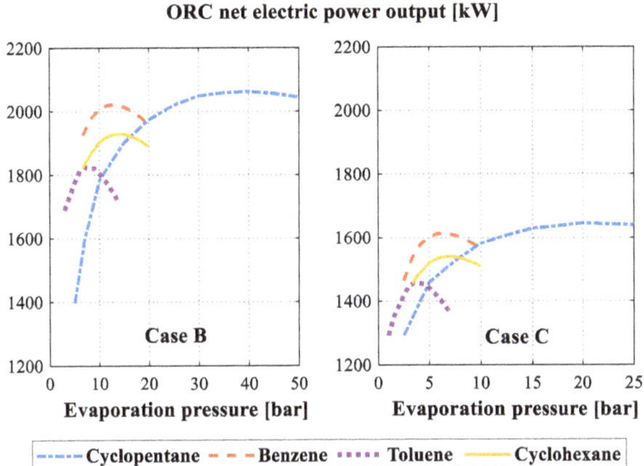

Figure 9. Sensitivity analysis results on organic fluid and IHTF design temperature: Case B and Case C layout.

4.2. Off-Design Analysis

Based on sensitivity analysis results and design assumptions of Tables 4–6, off-design evaluation of proposed layouts was carried out, considering Cyclopentane as working fluid in all the cases. The evaporation pressure in design conditions was set equal to the optimum values, based on the design thermodynamic results described in paragraph 4.1. The input variables for the three analyzed layouts (see Table 1) were varied in order to simulate the ORC performance under variable DH heat demand.

Figure 10 shows the ORC net power output as a function of different considered input variables, namely the IHTF HE inlet/outlet gas temperature and inlet mass flow, for the different layout arrangements. The input variables ranged between the 110% design load condition and the minimum possible operating value. In detail, Figure 10a shows the performance behavior of Case A layout as a function of the temperature of the exhaust gas feeding the IHTF HE (T_{G2}, see Figure 6), for the different IHTF design temperature. The figure indicated that sizing the ORC on higher IHTF design temperature improved the power production in design conditions, but affected the performance in off-design and limited the operational range of the ORC. Indeed, in Case A1, the ORC could exploit exhaust gas temperature values down to 310 °C; for the Case A2, down to 260 °C; for the Case A3, down to 240 °C. Figure 10b shows the performance of Case B layout as a function of the gas outlet temperature (T_{G2}, see Figure 6), while, in Figure 10c, ORC power output was plotted versus the gas mass flow rate ($\dot{m}_{G1''}$, see Figure 6). In Case B, the exhaust gas outlet temperature ranged between 190 °C and 320 °C. In Case C, the ORC could operate with gas mass flow ranging between 12 and 36 kg/s.

Actually, an ORC could be regulated from full-load down to minimum 30% continuously with fast response, while it was able to increase its power output, up to 110% of its rated power, for a limited amount of hour. Thus, in this analysis, it was assumed that the ORC could work during the year even in the moments when the heat demand was lower than 3 MW, when the IHTF HE was fed with a thermal power higher than the design one until the ORC load did not exceed the 110%.

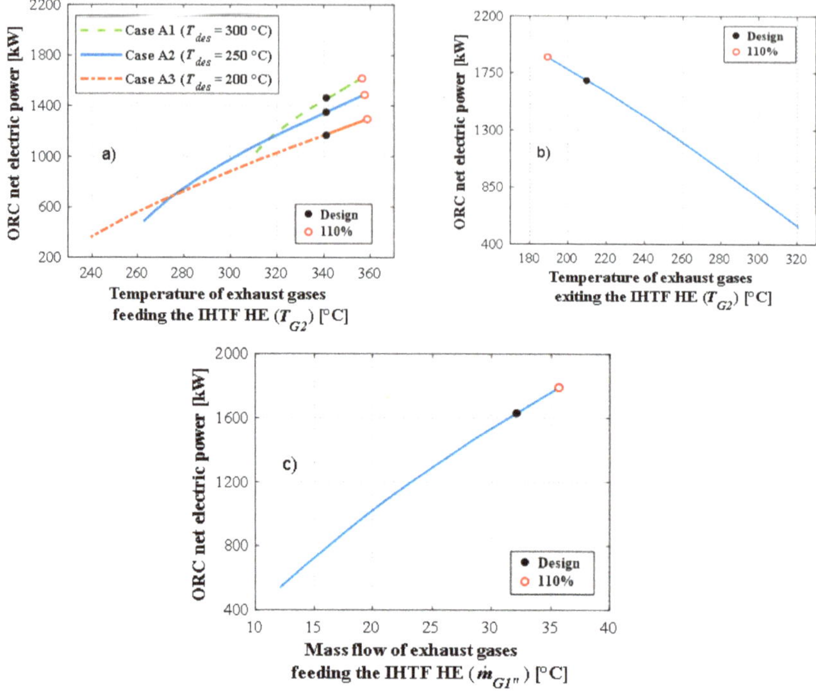

Figure 10. ORC net electric power at part-load conditions: (**a**) effect of gas temperature at IHTF HE inlet, for layout A and different T_{des} values; (**b**) effect of gas temperature at IHTF HE outlet, for layout B; (**c**) effect of gas mass flow.

4.3. Energy Evaluation

According to the DH yearly heat demand profile of Figure 2 and to the ORC performance trends of Figure 10, main comparative energy results were quantified and summarized in Table 7, showing the yearly fuel consumption and ICE generated energy, the ORC design and peak power output, the ORC average power during operating hours, the ORC yearly generated electrical energy, the ORC + ICEs yearly generated electrical energy, and additional generated energy, for each analyzed case study. The number of operating hours and generated electric power output for each investigated layout are also shown and compared in Figure 11, in the form of power outputs monotonic profiles. The ORC performance referred to the net ORC electric power output obtained as gross electric power output value minus IHTF and ORC pumps consumptions (corresponding to about the 8% of the ORC gross electric power output), minus air condenser fan consumptions (corresponding to another 8% of the ORC gross electric power output alone).

In the new arrangement, the ICEs worked the whole year at full-load, producing more energy than in the original set-up, in which they often operate at part-load conditions following the heat demand. In particular, in the new arrangement, the ICEs produced 170.89 GWh/years of additional electric energy, consuming 344.59 GWh/years of fuel more than in the original set-up. The exhaust gas was recovered, furthermore, to feed the IHTF HE, and an additional amount of energy was produced by means of the ORC. Therefore, the total electric energy produced by the new arrangement depended on the considered ORC layout performance.

Table 7. Main energy results for the analyzed configurations.

Comparative Results	Original Arrangement		New Arrangements	Difference	
Fuel consumption, F (GWh/year)	138.36		482.95	344.59	
ICE generated electric energy (GWh/year)	54.22		225.11	170.89	
Generated thermal energy (GWh/year)	50.05		50.05	0	
ORC Results	**CASE A1**	**CASE A2**	**CASE A3**	**CASE B**	**CASE C**
Design ORC power output (kW)	1465	1360	1183	1996	1627
Maximum ORC power output (kW)	1573	1496	1301	2027	1790
ORC average power during operating hours (kW)	1433	1177	988	1617	1451
ORC generated electric energy (GWh$_e$/year)	6.34	7.10	5.51	5.06	9.77
ORC + ICEs generated electric energy (GWh$_e$/year)	231.45	232.21	230.62	230.17	234.88
Additional generated electric energy, ΔE_{EL} [GWh$_e$/year] (a+b)	177.23	177.99	176.40	175.95	180.66
$GPES$ (%)	17.9	18.2	17.6	17.4	19.1

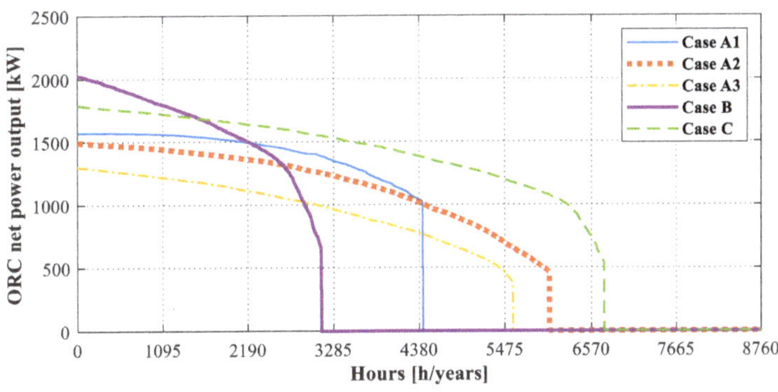

Figure 11. Power output monotonic profiles during the year for each investigated layout.

Case B showed the highest ORC power peak value, but not the highest amount of producible electric energy, due to the limited number of hours in which it operates during the year (about 3126). Thus, in order to evaluate the performance of the ORC on the year basis, it was fundamental to analyze its behavior also at part-load conditions and to take into account its regulation limits. In this case study, the number of operating hours turned out to be fundamental: the layout that allowed to produce the maximum amount of ORC electric energy per year, 9.77 GWh/year, was the Case C, case where the ORC could operate for the greatest number of hours, equal to 7050 h/year. More detailed results concerning the ORC best case (Case C) are reported in Appendix A for completeness. In particular, the monthly profile of heat demand, electric power generated by the old arrangement and the new one, and ORC net electric power output are grouped in Figure A1. Daily profile of heat demand and ORC net electric power output are presented and compared in Figure A2, for two days (February 15th and July 15th), chosen as representative of winter and summer operations of the DHN.

In order to measure the benefits of the ICE-ORC CHP production, a comparison was carried out between two scenarios, in terms of primary energy consumption: in the first scenario, the old CHP system was considered, while, in the second scenario, the new CHP was considered, where an additional electric energy production (ΔE_{EL}) occurred, and heat demand remained the same (see Table 7).

A relative global primary energy saving ($GPES$) could be calculated as follows:

$$GPES = \frac{F_{old} - F_{new}}{F_{old}} \tag{2}$$

where F_{old} is the primary energy consumption in the old scenario, and F_{new} is the primary energy consumption in the second one, respectively, calculated according to Equation (3) and Equation (4).

$$F_{new} = F_{ICEnew} + F_{BOILERnew} \tag{3}$$

$$F_{old} = F_{ICEold} + F_{BOILERold} + \frac{\Delta E_{EL}}{\eta_{EL\ GEN\ MIX} \cdot p} \tag{4}$$

In particular, the primary energy consumption could be split into three contributions: a contribution due to the ICEs operation, F_{ICE}, a contribution due to the boiler's operation, F_{BOILER}, and a third term due to, ΔE_{EL}. This term represented the fuel saved by producing ΔE_{EL} with the new arrangement in place of external electric energy production, and it was calculated considering the European electric generating mix. This latest contribution was estimated as the ratio between the additional generated electric energy from the engines and the ORC, ΔE_{EL}, and the average electric efficiency of the European electric generating mix, $\eta_{EL\ GEN\ MIX}$, and a coefficient, p, which took into account the grid losses depending on the feed-in voltage connection to the grid; in particular, p was intended to promote the feed-in of electricity at lower voltage.

In this analysis, $\eta_{EL\ GEN\ MIX}$ was assumed equal to 40%, which corresponded to the EU28 average electric efficiency of the electric generating mix in 2017 [25]. p was assumed equal to 0.985, which corresponded to the correction factor related to the avoided grid losses, for a CHP system connected to the grid at a voltage equal to 150 kV, as provided by the Directive on primary energy-saving calculation [26].

Results concerning the *GPES* are also reported in Table 7. The *GPES* analysis highlighted that the ICE-ORC arrangement introduced a positive primary energy saving. The *GPES*, however, differed for the different analyzed configuration; in particular, the highest value of the *GPES* was obtained for the Case C (19.1%), while the lowest value was obtained for the Case B (17.4%), for the aforementioned reasons. The inclusion of the ORC entailed positive values of *GPES* mainly because it allowed producing additional electric energy with remarkably high-efficiency values close to 52%, higher than the considered average European electric generating mix efficiency and in line with the current technology of mid-size combined cycles.

4.4. Techno-Economic Feasibility of Analyzed ICE-ORC Layouts

The ICE-ORC solution was compared to the original arrangement in terms of economic performance, by means of the differential net present value index. The differential net present value is defined in Equation (5), where ΔC_i is the differential cost, ΔR_i is the differential revenue at the i-th year, M_a is a factor assumed equal to 0.9 [27] that accounts for the ORC yearly maintenance costs, q is the discount rate assumed equal to 6%, and I_{ORC} is the investment on the ORC system. The differential cost was evaluated as the product between the natural gas price and the differential primary energy consumption (Equation (6)). Different gas price values were considered in this analysis: a medium value assumed equal to 33 EUR/MWh, a high value equal to 44 EUR/MWh, and a low value equal to 22 EUR/MWh. These values corresponded, respectively, to the average, the highest, and the lowest natural gas price encountered in the European countries at the beginning of 2019 [28]). The differential revenue was estimated according to Equation (7) as the product between the electric energy sell price and the differential generated energy. The investment cost of ORC was quantified according to the trend line in Figure 12, where specific investment cost was plotted as a function of the ORC size, based on product data of an ORC market leader manufacturer [23].

Equation (8) could be manipulated in order to evaluate the electric energy price that guarantees the return on investment in a given time period, i.e., the payback period (*PB*), using Equation (5).

$$\Delta NPV = \sum_{i=1}^{n} \frac{\Delta R_i \cdot M_a - \Delta C_i}{(1+q)^i} - I_{ORC} \tag{5}$$

$$\Delta C_i = \Delta F \cdot C_{FUEL} \tag{6}$$

$$\Delta R_i = \Delta E_{EL} \cdot C_{EL} \tag{7}$$

$$C_{EL} = \left(\frac{I_{ORC}}{\sum_{i=1}^{PB} \frac{1}{(1+q)^i} \cdot PB} + \Delta C_i \right) \cdot \frac{1}{\Delta E_{EL} \cdot M_a} \tag{8}$$

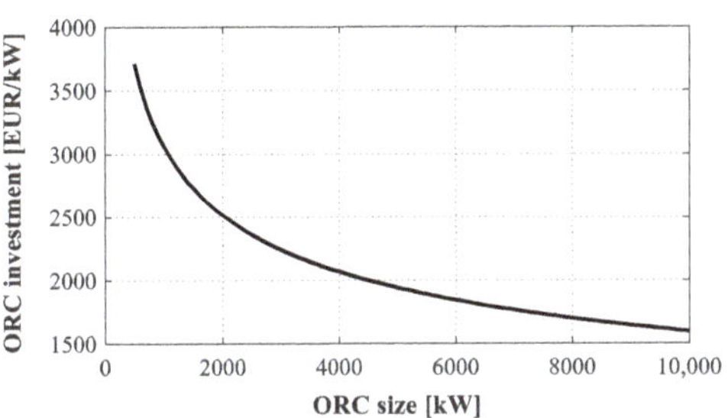

Figure 12. ORC specific investment cost as a function of plant size [23].

Figure 13 displays the results of the economic assessment in terms of electric energy sell price to return on investment in a given payback period. In particular, Figure 13a shows a comparison between the different ICE + ORC layouts, by considering the same natural gas price (the medium value), while Figure 13b shows the influence of the natural gas price. As expected, the payback period decreased when the electric energy price increased; indeed, higher electric energy prices led to higher revenues and, thus, to lower payback periods.

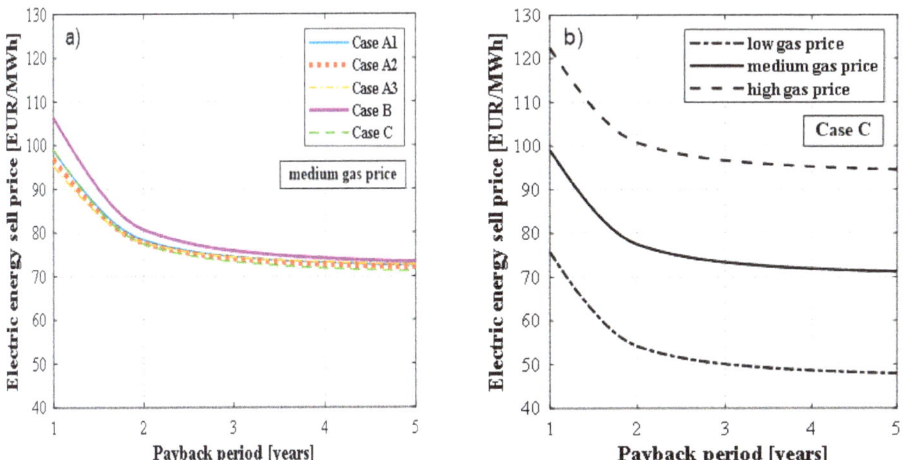

Figure 13. Electric energy sell price to return on the investment in a given payback period: (**a**) Comparison between the different ICE + ORC configurations; (**b**) Influence of the natural gas price value.

The trends were very similar for the different layouts, only Case B deviated slightly from the others. From the economic point of view, the Case B was the worst one because it presented the highest investment (due to the highest size) and the lowest yearly earning (due to the lowest yearly energy production). In general, the ORC solution proved to be a good investment since it allowed returning on the investment in barely 5 years, by selling the electric energy at a price of about 70 EUR/MWh, considering a medium natural gas price (33 EUR/MWh). When the higher natural gas price was considered, instead, the electric energy sell price could be lower in order to return on the investment in the same payback period; when lower natural gas price was considered, on the contrary, the electric energy sell price must increase. Especially for the Case C, in order to return on the investment in 5 years, the electric energy should be sold at a price of 95 EUR/MWh, considering the high gas price, or at a price of even 50 EUR/MWh, considering the low gas price.

5. Conclusions

This paper investigated the possibility of adding supplementary electric energy production to a CHP system at a service of a DHN, by converting the prime movers' exhaust heat with an ORC. The inclusion of the ORC allowed operating the prime movers at full-load (thus at their maximum efficiency), regardless of the heat demand, without dissipating not required high enthalpy-heat. Indeed, discharged heat was recovered by the ORC to produce additional electric power at high efficiency. In this work, a specific case study of an existing CHP system was analyzed; however, the same presented approach and similar considerations could be applied to any CHP feeding a DHN as a thermal user.

In the original arrangement, three internal combustion engines covered the heat demand together with the back-up boilers. ICEs load regulation strategy was determined in order to fulfill the DH demand while minimizing the heat dissipation to the ambient. The CHP in its original arrangement was compared to a new arrangement, including an ORC. In the new set-up, the ICEs worked at full-load for the whole year. The ORC was conceived to exploit the residual heat of the internal combustion engine exhaust gases, which has not been used by the district heating in part-load conditions. Three distinct layouts were investigated in this work: i) the ORC placed downstream the heat exchanger feeding the DH network; ii) the ORC supposed to be upstream the DH heat exchanger; iii) a parallel arrangement.

A sensitivity analysis of organic fluid and key cycle parameters was performed in order to identify, for each proposed arrangement, the optimum design of the ORC. Results showed that, for each analyzed configuration, Cyclopentane achieved the best performance compared to other hydrocarbons. The value of the optimal evaporation pressure depended on the considered configuration. As expected, the highest ORC performance was achieved for the highest value of the intermediate heat-transfer fluid circuit design temperature. Maximum power output (2027 kW) was achieved for the second layout solution, with an evaporation pressure of 40 bar and IHTF design temperature equal to 300 °C. In this case, the IHTF heat exchanger was fed with the exhaust gas at its highest temperature and mass flow, if compared with the other case studies. Based on sensitivity analysis results, the CHP-ORC hybrid plants were simulated by means of the commercial software *Thermoflex™* in design and in off-design conditions to account for ORC and DH part-load operation during the year.

The original optimal operation of the CHP plant was compared with the modified one in order to estimate, for each proposed integrated ICEs-ORC arrangement, the amount of additional generated electric power and the corresponding increase in primary energy consumption. The additional electrical energy produced during the year, thanks to the ORC, its operating hours, and total investment costs were quantified and compared among analyzed layouts. Results showed that the performance of the ORC, on the year basis, strongly depended on its part-load behavior and on its regulation limits. Indeed, even if the second layout exhibited the highest ORC power peak value, in design conditions, it did not provide the highest amount of producible electric energy on the year basis because of the limited number of hours in which it operated during the year. The layout that allowed to produce the maximum amount of ORC electric energy per year, 9.77 GWh/year, was the third one, i.e., the parallel arrangement. In the third set-up, the ORC peak power was lower than in the second (1790 kW), but it could operate for the greatest number of hours, namely 7050 h/year. Thus, the actual number of operating hours during the year, determined by the ORC regulation limits, turned out to be fundamental in the plant arrangement decision. An energetic index accounting for the global primary energy saving was introduced, in order to evaluate the convenience in producing the additional electric power with the modified CHP plant rather than with the original plant with the help of the electric generating mix. The energetic analysis demonstrated that all the proposed solutions granted to reduce the global primary energy consumption by about 18%. The best ICE + ORC solution, in particular, introduced a positive global primary energy saving equal to 19.1%. The inclusion of the ORC entailed significant savings in primary energy consumption because it allowed producing electric energy with an efficiency close to 52%, higher than the electric generating mix efficiency.

Finally, the economic feasibility of each CHP-ORC hybrid system layout was evaluated. The differential cost between the original and the new arrangement was related to the additional primary energy consumption. The differential revenue, instead, was proportional to the additional generated energy. The economic analysis showed that the second layout was the worst solution since it presented the highest investment (due to the highest size) and the lowest yearly earning (due to the lowest yearly energy production). Nevertheless, all the proposed ORC solutions proved to be a good investment since they allowed to return on the investment in barely 5 years, by selling the electric energy at a price equal to 70 EUR/MWh, considering a medium natural gas price equal to 33 EUR/MWh. When the higher natural gas price was considered, instead, the electric energy sell price could be lower in order to return on the investment in the same payback period; when lower natural gas price was considered, on the contrary, the electric energy sell price must increase. Especially for the best ICE+ORC case, in order to return on the investment in 5 years, the electric energy could be sold at a price of 95 EUR/MWh, considering the high gas price, or at a price of even 50 EUR/MWh, considering the low gas price.

In conclusion, results of this work allowed to state that the ORC technology effectively represents an interesting solution to make the plant regulation more flexible, with the purpose of optimizing economic revenue and energetic performance of the CHP plant, while not dissipating high-enthalpy valuable heat to the ambient. Given the promising results of this work, future studies will be dedicated to investigating the dynamic behavior of the DHN, by considering also thermal storage, in order to evaluate the possibility of managing the DHN in a more active and profitable way.

Author Contributions: Conceptualization, L.B. and F.M.; Data curation, N.T.; Funding acquisition, F.M.; Methodology, L.B. and F.M.; Project administration, A.D.P.; Software, N.T.; Supervision, L.B., A.D.P., and F.M.; Validation, A.D.P.; Writing—original draft, N.T.; Writing—review and editing, A.D.P. All authors have read and agreed to the published version of the manuscript.

Funding: This research received no external funding.

Conflicts of Interest: The authors declare no conflict of interest.

Nomenclature

Acronyms

CHP	Combined heat and power
DH	District heating
DHN	District heating Network
GR	Group
GPES	Global primary energy saving
HE	Heat exchanger
ICE	Internal combustion engine
IHTF	Intermediate heat transfer fluid
ORC	Organic Rankine cycle
WHR	Waste heat recovery

Symbols

Δ	Differential
C	Cost (EUR)
E	Energy (GWh)
f	Function of (-)
I	Investment costs (EUR)
M_a	Maintenance cost factor (-)
$\dot{m}$	Mass flow (kg/s)

n	Plant operating life (years)
η	Efficiency (%)
PB	Payback period (years)
q	Discount rate (%)
$\dot{Q}$	Thermal power (W)
R	Revenue (EUR)
T	Temperature (°C)
Subscripts	
des	Design
diss	Dissipated
el	Electric
exhaust	Exhaust
fuel	Fuel
max	Maximum
new	New
old	Old
th	Thermal

Appendix A

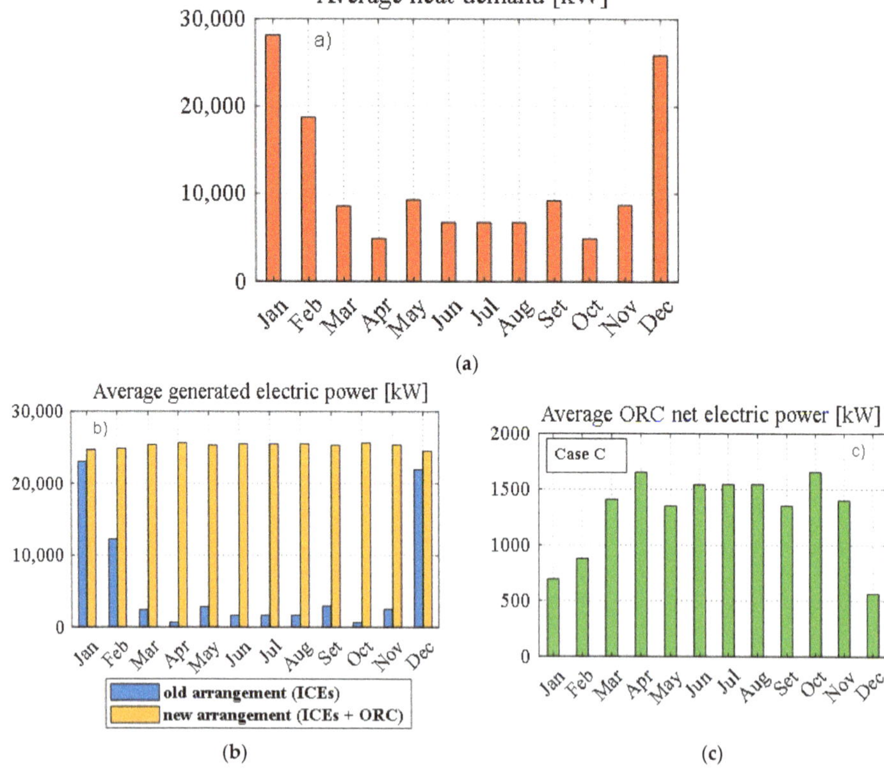

(a)

(b)

(c)

Figure A1. Monthly profile of: (**a**) Thermal power demand; (**b**) Electric power generated by the old arrangement and the new one; (**c**) ORC net electric power output.

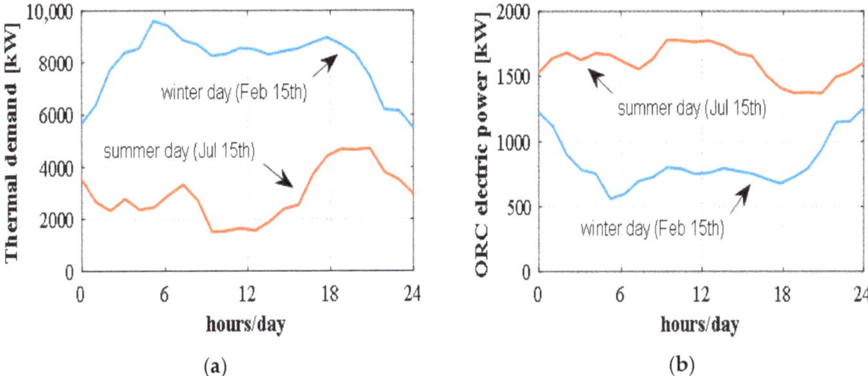

Figure A2. Daily profile of Thermal demand and ORC net electric power output for two representative days: (**a**) 15th of February; (**b**) 15th of July.

References

1. European Commission. Energy Efficiency Directive. 2019. Available online: https://ec.europa.eu/energy/en/topics/energy-efficiency/targets-directive-and-rules/energy-efficiency-directive (accessed on 1 December 2019).
2. European Commission. EU 2020 Target for Energy Efficiency. 2019. Available online: https://ec.europa.eu/energy/en/topics/energy-efficiency/targets-directive-and-rules/eu-targets-energy-efficiency (accessed on 1 December 2019).
3. Macchi, E.; Astolfi, M. *Organic Rankine Cycle (ORC) Power Systems—Technologies and Applications*; Woodhead Publishing: Duxford, UK, 2017.
4. Quoilin, S.; Van Den Broek, M.; Declaye, S.; Dewalle, P.; Lemort, V. Techno-economic survey of Organic Rankine Cycle (ORC) systems. *Renew. Sustain. Energy Rev.* **2013**, *22*, 168–186. [CrossRef]
5. Yang, K.; Zhang, H.; Song, S.; Yang, F.; Liu, H.; Zhao, G.; Zhang, J.; Yao, B. Effects of Degree of Superheat on the Running Performance of an Organic Rankine Cycle (ORC) Waste Heat Recovery System for Diesel Engines under Various Operating Conditions. *Energies* **2014**, *7*, 2123–2145. [CrossRef]
6. Han, Z.; Li, P.; Han, X.; Mei, Z.; Wang, Z. Thermo-Economic Performance Analysis of a Regenerative Superheating Organic Rankine Cycle for Waste Heat Recovery. *Energies* **2017**, *10*, 1593. [CrossRef]
7. Valencia, G.; Fontalvo, A.; Cárdenas, Y.; Duarte, J.; Isaza, C. Energy and Exergy Analysis of Different Exhaust Waste Heat Recovery Systems for Natural Gas Engine Based on ORC. *Energies* **2019**, *12*, 2378. [CrossRef]
8. Wang, X.; Tian, H.; Shu, G. Part-Load Performance Prediction and Operation Strategy Design of Organic Rankine Cycles with a Medium Cycle Used for Recovering Waste Heat from Gaseous Fuel Engines. *Energies* **2016**, *9*, 527. [CrossRef]
9. Song, S.; Zhang, H.; Zhao, R.; Meng, F.; Liu, H.; Wang, J.; Yao, B. Simulation and Performance Analysis of Organic Rankine Systems for Stationary Compressed Natural Gas Engine. *Energies* **2017**, *10*, 544. [CrossRef]
10. Yi, J.Y.; Kim, K.M.; Lee, J.; Oh, M.S. Exergy Analysis for Utilizing Latent Energy of Thermal Energy Storage System in District Heating. *Energies* **2019**, *12*, 1391. [CrossRef]
11. Arabkoohsara, A.; Namib, H. Thermodynamic and economic analyses of a hybrid waste-driven CHP–ORC plant with exhaust heat recovery. *Energy Convers. Manag.* **2019**, *187*, 512–522. [CrossRef]
12. Marty, F.; Serra, S.; Sochard, S.; Reneaume, J.M. Simultaneous optimization of the district heating network topology and the Organic Rankine Cycle sizing of a geothermal plant. *Energy* **2018**, *159*, 1060–1074. [CrossRef]
13. Ramireza, M.; Epeldea, M.; Gomez de Artechea, M.A.; Hammerschmidc, P.A.; Baresid, M.; Monti, N. Performance evaluation of an ORC unit integrated to a waste heat recovery system in a steel mill. *Energy Procedia* **2017**, *29*, 535–542. [CrossRef]
14. Grljušić, M.; Medica, V.; Radica, G. Calculation of Efficiencies of a Ship Power Plant Operating with Waste Heat Recovery through Combined Heat and Power Production. *Energies* **2015**, *8*, 4273–4299. [CrossRef]
15. *Thermoflex 27.0*; Thermoflow Inc.: Subdury, MA, USA, 2019.

16. Ancona, M.A.; Melino, F.; Peretto, A. An Optimization Procedure for District Heating Networks. *Energy Procedia* **2014**, *61*, 278–281. [CrossRef]

17. Doty, S.; Turner, C.W. *Energy Management Handbook*, 6th ed.; Taylor and Francis: Lilburn, GA, USA, 2006.

18. Ancona, M.A.; Baldi, F.; Bianchi, M.; Branchini, L.; Melino, F.; Peretto, A.; Rosati, J. Efficiency improvement on a cruise ship: Load allocation optimization. *Energy Convers. Manag.* **2018**, *164*, 42–58. [CrossRef]

19. Shi, L.; Shu, G.; Tian, H.; Deng, S. A review of modified Organic Rankine cycles (ORCs) for internal combustion engine waste heat recovery (ICE-WHR). *Renew. Sustain. Energy Rev.* **2018**, *92*, 95–110. [CrossRef]

20. Zhang, T.; Zhu, T.; An, W.; Song, X.; Liu, L.; Liu, H. Unsteady analysis of a bottoming Organic Rankine Cycle for exhaust heat recovery from an Internal Combustion Engine using Monte Carlo simulation. *Energy Convers. Manag.* **2016**, *124*, 357–368. [CrossRef]

21. Lemmon, E.W.; Bell, I.H.; Huber, M.L.; McLinden, M.O. *NIST Standard Reference Database 23: Reference Fluid Thermodynamic and Transport Properties-REFPROP, Version 10.0*; National Institute of Standards and Technology: Gaithersburg, MD, USA, 2010; Volume 22.

22. Therminol 62 Brochure. Available online: https://www.therminol.com/sites/therminol/files/documents/TF-8692_Therminol_62.pdf (accessed on 1 December 2019).

23. Bianchi, M.; Branchini, L.; De Pascale, A.; Melino, F.; Peretto, A.; Archetti, D.; Campana, F.; Ferrari, T.; Rossetti, N. Feasibility of ORC application in natural gas compressor stations. *Energy* **2019**, *173*, 1–15. [CrossRef]

24. Bianchi, M.; Branchini, L.; De Pascale, A.; Melino, F.; Orlandini, V.; Peretto, A.; Archetti, D.; Campana, F.; Ferrari, T.; Rossetti, N. Energy recovery in natural gas compressor stations taking advantage of organic Rankine cycle: Preliminary design analysis. In Proceedings of the ASME Turbo Expo 2017, Charlotte, NC, USA, 26–30 June 2017.

25. Electric Generation Mix Efficiency Data. *Fattori di Emissioni Atmosferica di Gas a Effetto Serra Nel Settore Elettrico Nazionale e Nei Principali Paesi Europei*; ISPRA: Rome, Italy, 2019; p. 69.

26. Correction Factor Related to the Losses Avoided Along the Grid, for a Chp System. Data Available from Italian Ministerial Decree Related to Combined Heat and Power Primary Energy Saving Calculation: Allegato VII del DM 4 Agosto 2011 GSE—Cogenerazione ad Alto Rendimento. Available online: https://www.gazzettaufficiale.it/eli/id/2011/09/19/11A12046/sg (accessed on 1 December 2019).

27. Bianchi, M.; Branchini, L.; De Pascale, A.; Melino, F.; Orlandini, V.; Peretto, A.; Archetti, D.; Campana, F.; Ferrari, T.; Rossetti, N. Techno-Economic Analysis of ORC in Gas Compression Stations Taking into Account Actual Operating Conditions. *Energy Procedia* **2017**, *129*, 543–550. [CrossRef]

28. Eurostat. Natural Gas Price Statistics. 2019. Available online: https://ec.europa.eu/eurostat/statistics-explained/index.php/Natural_gas_price_statistics#Natural_gas_prices_for_non-household_consumers (accessed on 1 December 2019).

Article

Pressure Pulsation and Cavitation Phenomena in a Micro-ORC System

Nicola Casari, Ettore Fadiga, Michele Pinelli, Saverio Randi and Alessio Suman *

Department of Engineering (DE), University of Ferrara, 44122 Ferrara, Italy; nicola.casari@unife.it (N.C.);
ettore.fadiga@unife.it (E.F.); michele.pinelli@unife.it (M.P.); saverio.randi@unife.it (S.R.)
* Correspondence: alessio.suman@unife.it

Received: 9 May 2019; Accepted: 5 June 2019; Published: 8 June 2019

Abstract: Micro-ORC systems are usually equipped with positive displacement machines such as expanders and pumps. The pumping system has to guarantee the mass flow rate and allows a pressure rise from the condensation to the evaporation pressure values. In addition, the pumping system supplies the organic fluid, characterized by pressure and temperature very close to the saturation. In this work, a CFD approach is developed to analyze from a novel point of view the behavior of the pumping system of a regenerative lab-scale micro-ORC system. In fact, starting from the liquid receiver, the entire flow path, up to the inlet section of the evaporator, has been numerically simulated (including the Coriolis flow meter installed between the receiver and the gear pump). A fluid dynamic analysis has been carried out by means of a transient simulation with a mesh morphing strategy in order to analyze the transient phenomena and the effects of pump operation. The analysis has shown how the accuracy of the mass flow rate measurement could be affected by the pump operation being installed in the same circuit branch. In addition, the results have shown how the cavitation phenomenon affects the pump and the ORC system operation compared to control system actions.

Keywords: micro-ORC; gear pump; CFD; mesh morphing; pressure pulsation; cavitation; dynamic analysis

---

1. Introduction

Over the years several micro-ORC systems have been developed [1] with the aim of understanding and improving the performance of different expander devices, pumping systems, and heat exchanger technology. Despite their effectiveness in energy recovery, micro-ORC systems are characterized by low-efficiency values, especially in cases where the net electric power is less than 10 kWe [2]. In these applications, positive displacement machines are commonly used for expanding and pumping processes [3]. Regarding the expander technology, the review by Bao and Zhao [4] pointed out that, for micro energy systems, the volumetric expander technology is the most preferable in terms of efficiency and cost. In particular, scroll, screw, and rotary vane expanders are the most suitable for micro-ORC systems. When considering ORC systems of net electrical power output around 1 kWe, scroll and screw expanders are a well-known technology for energy conversion, but when the temperature of the hot source falls below 100 °C, other machines have to be used. For example, an innovative way to generate a small amount of electrical power (a few Watts) is to utilize free-piston expanders [5,6]. From the experimental side, the use of a positive displacement machine requires a specific test-rig to reproduce the actual operating conditions in terms of pressure, fluid characteristics and loads. Commonly, the presence of simplified prototypes and/or different working conditions could determine several differences between the actual machine performance and the measured ones. To help the experimental analysis, a lumped-parameter model is often used for extending the analysis to other operating conditions or other fluids [7,8], increasing the design capabilities and limiting the cost and time.

For the pump, no specific indication can be found in the literature, though a greater number of applications are equipped with a positive displacement machine, usually gear pumps. This widespread technology is common adopted for pressurizing the refrigerant (in most cases a certain percentage of oil mass fraction is added to the ORC systems to ensure the proper lubrication of the positive displacement machine) even if the gear pump operation is affected by two main issues, related to the (i) pressure pulsations and (ii) cavitation phenomenon.

Gear pumps are a class of robust positive displacement pumps that can work on a wide range of pressures and rotational speeds, but they are responsible for noise and vibration phenomena that could affect the circuit in which they operate. The damping effects provided by the pipeline are usually not able to eliminate the vibrations [9,10], which are also responsible for disturbances and, sometimes, the damage of measurement devices. Pulsations are due to the pressure ripples provided by the pump, which are driven by the engagement process and the position of the relief grooves [11].

In several applications, micro-ORC systems are equipped with temperature probes, pressure transducers and flow meters. These devices are used to characterize the system and, in addition, for controlling its operating point when external conditions change, such as the electric load and the heat and sink temperature values. Noise and vibrations could, in turn, affect their operations and increase the uncertainty of the measuring process. In particular, the Coriolis flow meter (which is one of the most used technology in micro-ORC test benches) can be greatly affected due to its technology and measuring principle [12]. Despite their accuracy and measuring performance, several analyses have been carried out over the year demonstrating how these devices could be affected by vibrations and pressure pulsations [13,14].

In addition, gear pumps are a class of positive displacement machine that shows high sensitivity to cavitation phenomena, in particular as a function of rotational speed. Higher rotational speed leads to higher volumetric efficiency but, at the same time, the pressure losses and the narrow gaps could cause local pressure drops that could lead to cavitation [15,16]. In ORC applications, this phenomenon is even more detrimental due to the thermodynamic state of the liquid pumped. In fact, the refrigerant in this section is very close to the saturated liquid curve and, so, even small pressure variations could originate phase change (from liquid to vapor) at the pump suction section. Cavitation produce erosion issues and is responsible for the reduction of the mass flow rate supplied by the pump. In this case, the control unit of the ORC system has to detect this discrepancy between the expected mass flow rate and the actual one and react by increasing the pump rotational speed.

Given this background, it is easy to understand how the use of a gear pump in a micro-ORC system needs to be analyzed and evaluated according to both the performance and reliability points of view. Pressure pulsations and cavitation affect the ORC system operation and the entire control system. Since these phenomena are mainly due to the pump operation, in this paper a time-dependent computational fluid dynamics (CFD) approach is proposed. The simulation strategy is based on a dynamic mesh approach, able to take into account the rotation and the engagement of the wheels, as well as the transient phenomena that affect the entire ORC. The dynamic mesh approach has been proven to be suitable for the analysis of such a machine, for discovering cavitation issues [17,18] and for calculating the performance [19,20].

The virtual model of the micro-ORC system is taken from an actual installation, described entirely by Bianchi et al. [21]. The capabilities of CFD simulations for analyzing the ORC performance are proven [22,23]; however, in this work, a step forward on the simulation and explanations of such phenomena is performed. The criteria adopted in this paper are defined by the Whole ORC Model (WOM) [22,23], which is a new strategy for improving knowledge in micro-scale energy systems. The WOM strategy concept arises from evidence that the performance of every single device is affected by the operation of the other components/devices that operate in the same system. Including, in the same virtual CFD model, a higher number of components allows for an analysis of their interaction and adjusting the control logic, which can be an effective strategy for increasing system reliability.

Aim of the Paper and Novelty

In this paper, a detailed analysis of pressure pulsation and cavitation phenomenon is carried out by means of transient CFD analysis. The numerical domain considered in the present analysis is depicted in Figure 1 and corresponds to the virtual model of a portion of a real ORC system [21]. As can be seen, starting from the liquid receiver (tank) (which collects the condensed liquid at the condenser outlet), the liquid passes through the Coriolis flow meter (CFM) (which is entirely modeled as reported in the internal view) and is processed by the pump. The pump is installed in the lower part of the circuit and is coupled with the system through an upstream and a downstream valve. The liquid refrigerant is then piped into the regenerator, which has the task of preheating it by removing heat from the refrigerant at the expander outlet. The virtual model depicts the actual installation in a very detailed form: bends, valves and tube-fittings are virtualized, generating a virtual test-bench. The innovative approach for studying the ORC operation reported in the present work is based on the application of a fully 3D transient numerical simulation of the entire system and allows us to analyze:

- the effect of gear pump operation (in terms of pressure pulsation) on the CFM measurement performance. With the dynamic analysis it is possible to extract information about the frequency and magnitude of the vibrations, which can affect the CFM due to the gear pump operation;
- the effect on the pump operation of the actual system layout and design. The virtualization of the ORC systems, which includes valves, bends, tank, regenerator, and measuring device (such as the CFM) allow for the determination of the actual flow conditions (in terms of velocity and pressure) at the inlet port of the gear pump;
- the effect of cavitation on the pump operating points. By coupling the control system strategy and the CFD analysis, it is possible to detect how the cavitation phenomenon affects the ORC operation and which solution can be adopted for improving the control capability of the system, with the aim of preserving the performance and the reliability of each component.

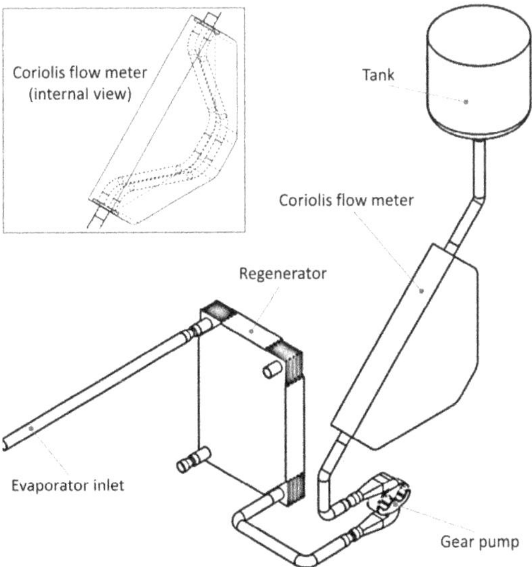

Figure 1. The virtual model of the micro-ORC system: tank, Coriolis flow meter, gear pump, and regenerator.

2. Materials and Methods

As already pointed out, the virtual model of the ORC system is realized with reference to an actual installation. The operating conditions are taken directly from the experimental data described in [21]. The experimental data are used for (i) validating the numerical results and (ii) feeding the numerical model with actual operating conditions, allowing the analysis of specific fluid dynamic effects (cavitation at the gear pump inlet and pressure pulsation on the CFM due to the gear pump operation), which can be difficult to detect by means of an experimental approach.

The prototypal micro-ORC energy system used as a reference was specifically designed for in-field operations but, at the same time, was equipped with several additional sensors/probes in order to detect its operating points and performance. The ORC system operates with R134a as a working fluid and its main components are as follows: (i) a brazed plate heat exchanger with 64 plates as evaporator, (ii) a brazed plate heat exchanger with 16 plates as a regenerator, (iii) a shell and tube heat exchanger as condenser, (iv) a volumetric three-piston radial engine used as expander, and finally, (v) a volumetric gear pump controlled by an inverter that supplies the organic fluid over the ORC system position under the liquid receiver (tank), realizing a column of water about 1 m high.

2.1. Computational Grid and Dynamic Mesh Approach

The numerical simulations and meshing phase are developed by means of the software Pumplinx 4.6.4 by Simerics Inc.® (Bellevue, WA, USA). The virtual model depicted in Figure 1 is discretized by means of a Cartesian grid (binary-tree mesh type) for the stationary domains (such as piping and gear pump body) and a body-fitted curvilinear mesh type for the rotary domains (such as the gear pump wheels). Figure 2 reports two mesh details of the stationary components: the outlet section of the regenerator and the inlet section of the CFM. The stationary mesh is composed of about 3 million elements. Finer grid elements are adopted at the variation of the passage area, especially in the proximity of the inlet and exit valves. Due to these geometric features, the element size of the mesh is adapted according to the passage area. In particular, in each section, an element number in the range of 20–30 was adopted along the pipe diameter. A detailed description of the element size adopted for the gap discretization is reported in the following section.

The dynamic mesh approach was adopted with the aim of analyzing the transient phenomenon generated by the gear pump operation. For this reason, particular attention was given to the mesh generation of this component. The internal volume of the gear pump has been discretized through a structured mesh able to account for the wheel rotation and engagement over the transient analysis. The rotation and engagement processes determine the modification of the internal shape of the computational domain at each time-step. The computational mesh of this section consists of about 300,000 hexahedral elements, which have been associated with a deformation algorithm able to simulate the rotation of the wheels and their engagement. Figure 3 shows four successive instants related to the rotation and engaging of the gear pump wheels. It can be seen how the grid elements (in the vane) are able to adapt and change their shape, to accommodate the presence of the tooth of the opposite wheel. The rotating domain is connected to the suction and discharge ports through sliding interfaces able to couple the stationary and rotating domains. The subdivision between stationary and rotating domains inside the gear pump is depicted in Figure 4. The suction and delivery ports of the gear pump are connected via sliding interfaces to the rotating domain.

The computational mesh, composed of about 3 million elements and depicted in Figures 2–4, is the result of a mesh sensitivity analysis (from 2 million to 4 million elements) that was carried out with the aim of validating the numerical results against the experimental data. A detailed comparison and quantification of the deviation between the numerical results and the actual ORC operation is reported in the following sections.

2.2. Gear Pump Virtual Model and Main Features

The pump displacement is equal to 7.5×10^{-6} m^3. The volumetric flow rate provided by the gear pump can be adjusted according to the ORC requirements by varying the rotational velocity. This ideal displacement is affected by the unavoidable leakages that characterize these machines. By means of the present simulation strategy, it is possible to include in the fluid dynamic analysis all the leakages through the gaps: between the wheels and the pump casing (i.e., tooth flank—pump casing sides and tooth head—pump casing periphery) and the engagement region.

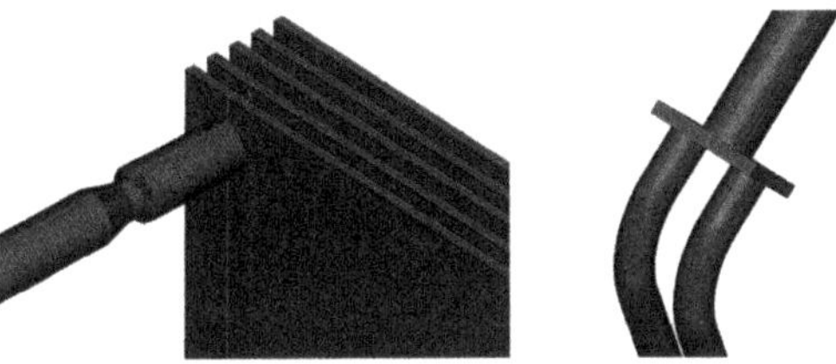

Figure 2. The unstructured Cartesian grid on stationary components: the outlet section of the regenerator and the CFM inlet section.

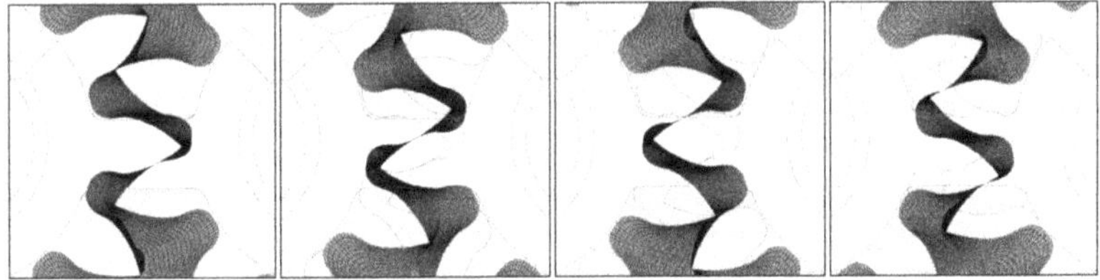

Figure 3. Subsequent instants of engagement process discretized by means of a dynamic mesh strategy.

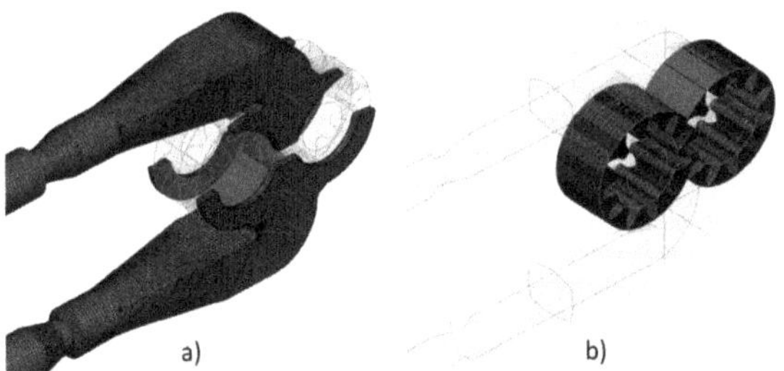

Figure 4. The computational mesh of the gear pump: (**a**) suction and delivery ports, and (**b**) rotating wheels equipped with hexahedral mesh elements.

The liquid flows through these gaps according to the pressure gradient, from the discharge port to the suction port determining a volumetric efficiency lower than one. This phenomenon characterizes all rotating volumetric machines and is one of the main points of interest for the fluid dynamics simulation of this type of machine. In fact, an ideal positive displacement machine has to perfectly seal the suction port with respect to the discharge port, but the manufacturing processes and the necessity to reduce the friction between the moving and fixed parts determine the axial and radial gaps.

The axial gaps are due to the axial clearance that exists between the wheels and the pump case. In the present analysis, the axial gap is equal to 20 μm and contributes to a decrease in the volumetric efficiency of the gear pump. In Figure 5a, the computational grid in the axial gaps is reported. It is realized by means of five layers of hexahedral elements and coupled with the rotating domain through a sling interface.

In a different way of the axial gaps, the clearance between the tooth head and the pump casing is a function of the angular position of the wheel and they vary in the range of 50–100 μm. Each angular position of the wheel is characterized by a different distance between the tooth head and the pump casing due to the eccentric position of the wheel. The eccentric position is related to the unbalanced pressure forces between the wheel region facing the delivery port (high pressure) and the wheel region facing the suction port (low pressure). The fluid pressure at the port acts on the overlocked wheel portion, generating a certain load. The fluid pressure at the delivery port is higher than the pressure value at the suction port; for this reason, the wheel is not concentric with respect to the pump casing but is located in an eccentric position responsible for the lower values for the tooth-head gap in the proximity of the suction port.

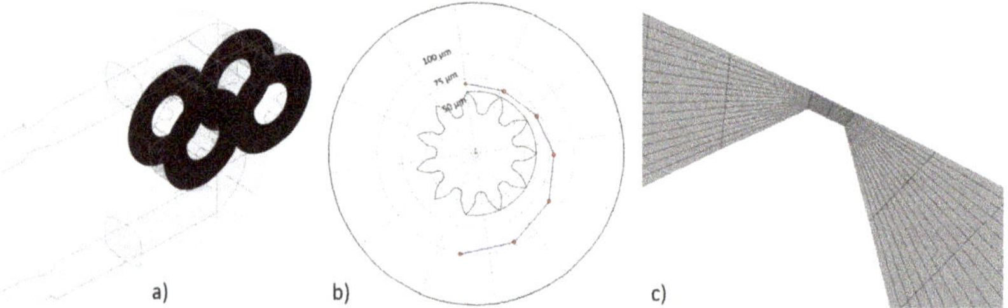

Figure 5. Particular domain discretization in correspondence of the gaps: (**a**) between the tooth flank and pump casing, (**b**) between the tooth head and the pump casing as a function of the angular position, and (**c**) computational mesh in the tooth head-case gap.

In Figure 5b the magnitude of the gaps is reported by means of a polar diagram, in which the delivery section is in the lower part while the suction section is in the upper part. They vary (in particular, increase) starting from about 70 μm up to 100 μm.

In the tooth-head gap, thanks to the deformation strategy adopted for the computational mesh, the number of elements used to discretize that region is identical to the number of elements used to discretize the vane volume. In particular, 16 elements were adopted for the radial discretization of the gear pump vane, that allows a proper resolution of the flow field. Figure 5c reports the computational mesh in correspondence of the gap between the tooth head and the pump case and shows the grid refinement in this particular region.

The last contribution to the reduction of volumetric efficiency is due to the engaging process. In the engaging region, when two teeth pairs come into contact, a trapped volume could arise and, in the case of the wrong design, it can experience a progressive volume reduction, leading to a sudden change in pressure [11]. To avoid this, the trapped volume is always connected with the high- or low-pressure chamber. This role is performed by purpose-built pockets in the lateral side of the pump casing, named relief grooves, whose dimensions are very important to the resulting dynamic behavior. The shape and position of this relief groove are depicted in Figure 6. The radial width of the relief groove must be balanced in order to (i) avoid the pressure rise caused by the trapped volume and (ii) limit the flow escaping from the delivery (high pressure) to the suction (low pressure) ports. As can be seen from the two successive instants shown in Figure 6, the relief groove is firstly connected to the suction (upper) and discharge (lower) ports (see Figure 6a) and then sealed by the contact between the two tooth pairs in contact (see Figure 6b). In the latter position, the teeth in contact do not allow fluid flow from the delivery to the suction port.

In addition to the negative contribution to the volumetric efficiency, the backflow from the discharge to the suction port through the relief groove also determines pressure pulsations that could affect the flow field at the suction ports of the gear pump. In this region, a high-pressure jet of fluid coming from the groove experiences a sudden pressure drop increasing the pressure variation and the cavitation effects. This phenomenon, strongly related to the transient phenomena experienced by the gear pump, will be considered in the subsequent analysis of the results, with particular attention to the repercussions of this effect on cavitation and on the variation in pressure field induced by the CFM, which is installed upstream of (but close to) the gear pump.

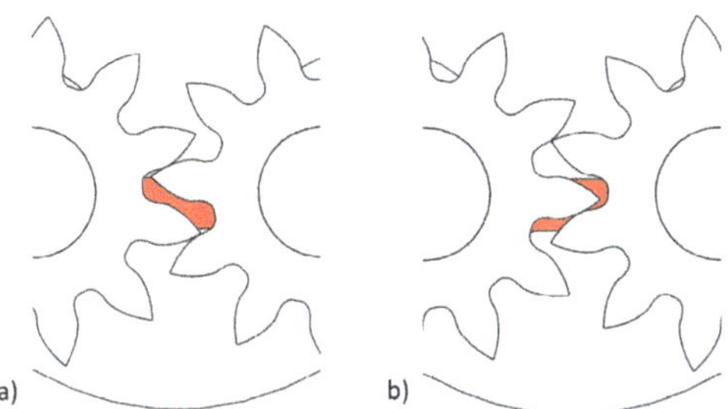

Figure 6. Engaging process compared to the relief groove shape: (**a**) the two pairs of teeth do not seal the relief groove with respect to the suction/discharge ports and (**b**) the teeth contact isolates the relief groove and seals the suction/discharge ports.

2.3. Operating Conditions

The refrigerant (in the liquid phase) considered in the present analysis is the R134a (F_3C-CH_2-F). Over the modeled flow path, the liquid refrigerant experience the pressure rise provided by the gear pump. The aim of the present analysis is to show dynamic phenomena such as pressure pulsation and cavitation, and for this reason, the present analysis is carried out avoiding the thermal power exchanged at the regenerator and the heat dissipated with the environment.

Dealing with dynamic analysis, the boundary conditions of the present investigation are averaged values according to the gear pump rotation. The transient analysis is carried out adopting a time-step adequate to solve the 0.5° rotation of the gear wheels.

The liquid receiver (tank) is considered pressurized at a fixed pressure level of 601,325 Pa, while at the evaporator inlet, the imposed pressure is 1,601,325 Pa. These pressure boundary conditions are in line with the operating point reported in [21], and are bounded by the heat source and sink constituted of water, having a temperature in the range 18–25 °C at the condenser and 60–90 °C at the evaporator.

According to the pressure boundary condition, the gear pump rotational speed is adjusted (in the range from 450 rpm to 475 rpm) in order to guarantee a mass flow rate of about 0.1 kg/s, which is considered the target value for the present analysis. Since the pump operation is affected by the leakages and the cavitation phenomena, the target mass flow rate is guaranteed with different rotation speed related to the sub-cooling degree that occurs at the condenser. The sub-cooling degree determines the liquid saturation pressure, which represents the most important values for the cavitation phenomenon. Starting from a reference condition (which corresponds to 450 rpm, liquid temperature of 293 K and saturation pressure of 569,060 Pa), two additional operating points have been simulated according to the working parameter of the ORC control system (variation of the liquid temperature and pump rotational speed). Details about these two additional analyses will be reported later.

As described above, the suction section of the pump is affected by high-speed gradients of the fluid velocity due to the simultaneous presence of (i) wheel rotation and (i) backflow through the relief groove from the engaging region. These phenomena, associated with the thermodynamic condition of the liquid phase at the pump inlet (close to the saturated condition), and pressure losses introduced by the piping and fittings (pipes, valves, CFM, and bends) make the cavitation phenomenon one of the main issues of the present application (gear pump in a micro-ORC system).

2.4. Model Setup and Validation

With reference to the previous description, all the numerical simulations were carried out by imposing a constant static pressure at the inlet (601,325 Pa) and at the outlet (1,601,325 Pa) sections of the virtual model depicted in Figure 1. For turbulence modeling, the k–ε turbulence model was adopted. A constant liquid density (1226 kg/m^3) and viscosity (2.02 × 10^{-4} Pa s) values are also specified for each operating condition. The cavitation phenomenon is modeled by means of the equilibrium dissolve gas model implemented in Pumplinx 4.6.4 [24].

As mentioned, the numerical model was validated against experimental data taken from [21]. The sensitivity analysis of the computational mesh was carried out with the aim of reproducing the experimental results and, at the same time, limiting the computational effort. Figure 7 reports the comparison between the mass flow rate m obtained from the experimental data and from the numerical solutions with respect to the gear pump rotational speed ω. The mass flow rate values are taken as an averaged value of the instantaneous mass flow rate provided by the pump. The data are in reasonable agreement (error bars equal to 10%), especially in the range from 425 rpm to 525 rpm. For cases with lower values of rotational speed, the presence of a lubricant oil (in a fraction between 3% and 6%, see [21]) allows the sealing process in the flank and radial gaps and, for this reason, the mass flow rate looks greater than the values obtained in numerical simulations.

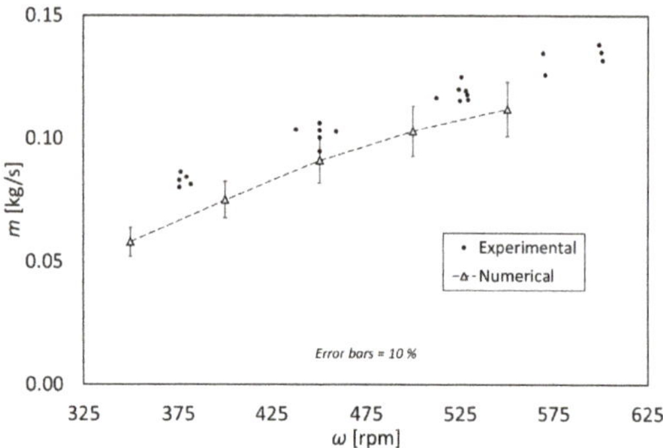

Figure 7. Model validation: comparison between the experimental data (taken from Bianchi et al. [21]) and the numerical results.

3. Results: Pump Operation

In this section, the results of the transient fluid dynamic analysis carried out with the dynamic mesh strategy will be reported. The results have to be referred to the angular position of the wheel. Each wheel is composed of 11 teeth, and for this reason, each plot will refer to a 32.7° rotation step. The reference of the angular position a (which corresponds to the origin of each plot) will be superimposed on each graph.

3.1. Mass Flow Rate and Velocity Field

The simulation strategy employed allows us to know the flow rate supplied by the positive displacement machine according to the fluid characteristics, rotational speed and operating conditions of the plant (pressure difference). The trend reported in Figure 8 shows the mass flow rate at the outlet section of the considered ORC virtual model (evaporator inlet). The mass flow is not constant but the trend follows the operation of the gear pump, which releases a certain volume of fluid in relation to its angular position. The peak of the mass flow rate corresponds to the vane released at the pump outlet section. Thanks to the time resolution of the transient simulation, it is possible to calculate, with a good approximation, the average flow rate processed by the gear pump during its rotation. For the actual rotational speed (450 rpm), the average value of the mass flow rate is equal to 0.091 kg/s, which is the result of the contemporary presence of the subsequent vane released at the discharge section and leakage driven by the pressure difference from the outlet to the suction port.

The average value of mass flow rate is affected by the leakages that occur in the pump region, reported and described in the previous section. Fluid dynamic representations of the flow leakages are reported in Figure 9. From the high-pressure section (pump discharge section), the liquid refrigerant flows through the axial gap (Figure 9a) and through the tooth-gap (Figure 9b), reducing the volumetric efficiency of the gear pump. Therefore, these leakages reduce the net flow rate processed by the pump measured at the outlet section since a small part of the fluid already processed by the wheels returns from the discharge to the suction port.

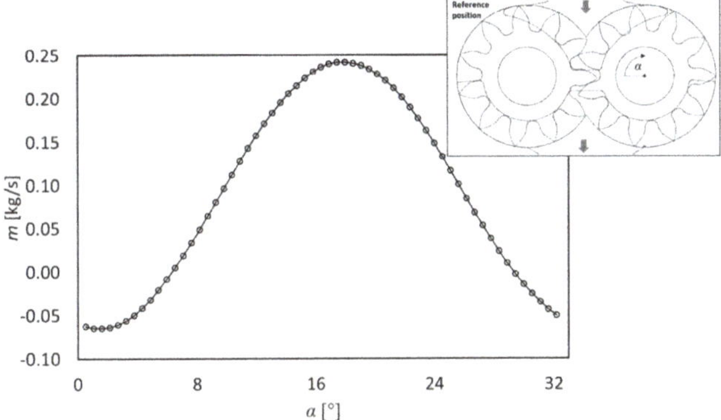

Figure 8. Mass flow rate trend according to the angular position of the wheel (see reference position).

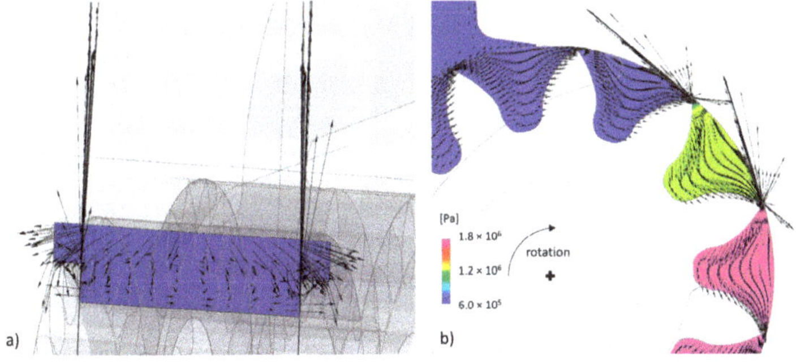

Figure 9. Velocity vector in the gaps: (**a**) tooth-flank gap and (**b**) tooth-head gap.

In Figure 9b, two families of vectors can be seen: the vectors with the same direction of wheel rotation versus vectors with opposite directions. The agreed vectors are those that represent the fluid moving from the suction to the delivery thanks to the sealing process provided by the teeth and thanks to the no-slip condition imposed on the wheel surfaces, while the discordant vectors represent the fluid that, driven by the pressure difference, flows from the delivery to the suction ports through the radial clearance (tooth-head gaps).

3.2. Pressure Field and Pulsations

In the present section, the pressure pulsations at the gear pump inlet and outlet ports are reported. The pressure pulsations were detected upstream and downstream of the gear pump thanks to the use of two virtual probes facing the suction and discharge ports. The positions of the two measurement points are shown in Figure 10. In these ports, the pump operation, characterized by the teeth passages in front of the ports, generates high turbulence and mixing. At the inlet section (p_{IN}), the fluid trapped in the meshing region determines the backflow of fluid, which is drawn from the delivery and released towards the suction port. According to the angular position of the wheel, the pressure reduction experienced by the pressure probe at the inlet section is due to the vane-filling process that occurs for the vane that finishes the engagement process, generating a sudden increment of the available volume at the suction section. At the outlet section (p_{OUT}), the pulsations are mainly due to the delivery of the trapped volume between two consecutive teeth. The results of all these phenomena on the suction and discharge ports are shown in Figure 10. Figure 10 also shows the sealing process and the consequent pressure field referred to a mid-plane of the tooth width. The pressure increment involves two consecutive vanes in the upper part of the pump (close to the suction section), according to the gaps in radial distribution shown in Figure 5b.

From the mass flow trend reported in Figure 8, it can be seen that, for a certain angular position range of 10° wide, a reverse flow condition characterizes the ORC system. Based on these findings, in Figure 11 two instants have been reported, representative of the flow conditions at 0° and 16.5°. These two conditions correspond to the instant at which the flow is reversed (0°) and the time of maximum flow (16.5°). The flow field is taken according to the section plane depicted in the figure located in the discharge section of the regenerator. The pressure pulsation implies that, for a certain interval, the flow moves from the outlet section towards the regenerator driven by the pressure difference. Figure 11a shows that the discharge pressure is greater than that generated by the pump, determining the reverse flow. By contrast, in Figure 11b, the pressurization generated by the gear pump is greater than what exists at the outlet, and the fluid is supplied to the regenerator outlet.

The pressure depicted in Figure 11 also highlights the pressure losses due to fittings and valves. The pressure loss reaches 2000 Pa, generated by the restriction in the flow passage area corresponding to the valve that divides the regenerator (highlighted in the sketch) and the evaporator (not considered in the present analysis) units. The flow passage section varies according to the valve and fitting internal geometries responsible for the localized pressure drop.

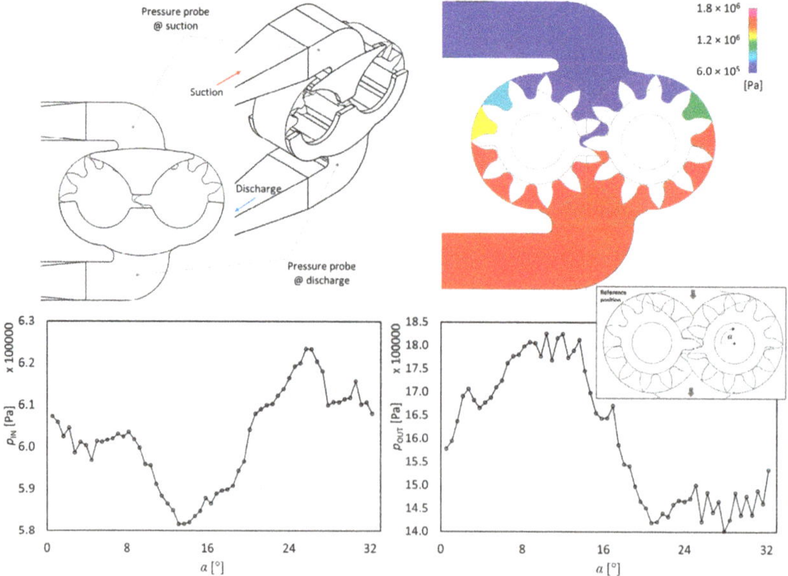

Figure 10. Pressure pulsation at the suction and discharge ports of the gear pumps (see reference position). The pressure field is referred to as the mid-plane of the tooth width.

4. Results: Response of the Cycle

4.1. Dynamic Effects on Coriolis Flow Meter

The pressure pulsation generated by the pump operation influences the entire ORC system. The CFM operating principle is based on the Coriolis force. The Coriolis force is due to the rotation. Each CFM includes one or more measuring tubes, which are equipped with an exciter able to generate a controlled oscillation. As soon as the fluid begins to flow inside the measuring tube, another oscillatory motion is imposed on the same tube due to the fluid inertia. Two sensors detect this change in the tube oscillation amplitude, frequency and phase difference. This difference is a direct measure of the mass flow rate. By taking into consideration the operating principle of a CFM, it is clear how a pressure pulsation may alter the measurement, introducing vibrations and oscillation not related to the mass flow rate. Based on these considerations, thanks to the transient analysis and the dynamic mesh approach, the pressure pulsations to which the CFM is subjected were monitored.

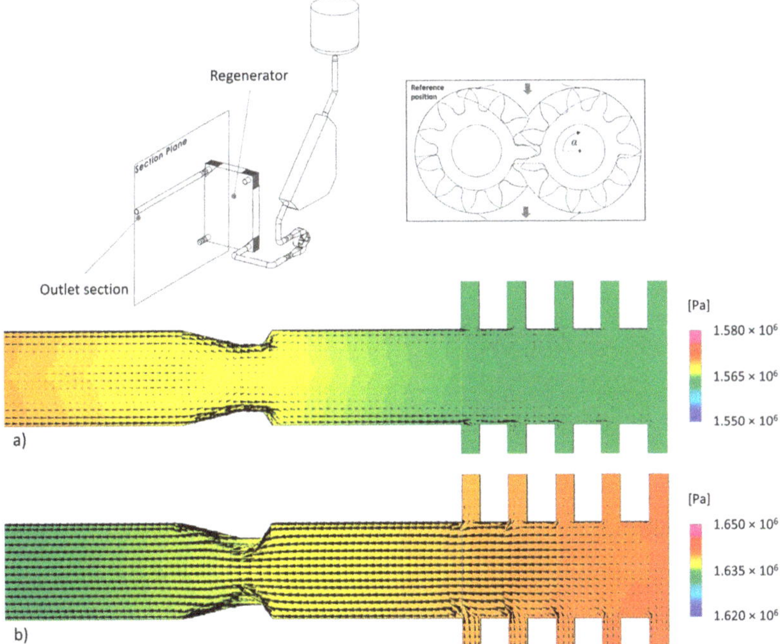

Figure 11. Pressure field and velocity vector at the regenerator outlet section at (**a**) 0° and (**b**) 16.5° (according to the reference position).

As described and reported in Figure 1, in the present ORC system, the CFM is installed before the gear pump, in line with the aim of reducing the effects due to the pressure pulsation generated by the gear pump. A different installation (CFM positioned after the gear pump) could determine higher values of pressure pulsation and, in turn, more detrimental effects due to flow disturbances on the CFM measuring performance. In many cases, the manufacturer of the flow meter indicates the proper distance between other devices, fittings and bend and the flow meter, but, considering the dynamic effects of positive displacement machine, the pressure pulsation involves the overall systems, implying that particular attention must be paid to the flow meter usage. The aim of this data post-processing is to show how, instead of proper installation, the interaction between two components could determine additional issues in the ORC control system.

The present analysis reports the pressure pulsation due to the gear pump generation without considering the damping effects provided by the frame that equipped the ORC systems (see, for example, the ORC skid in [21]). Therefore, the amplitude of the pulsations is intended as the greatest possible in terms of pressure and force. Also, the vibrational response of the structure is not considered in this work. Therefore, the analysis is intended to show the capabilities of the dynamic simulation to detect important features related to the system control beyond the mostly know results related to the flow field and machine performance. The inertia of fluid, together with the geometric features and pump operation work at the same time, and drive the interaction between two or more components. This type of analysis makes it possible to extract general guidelines for reducing the negative effects of pump operation on the CFM measurement performance.

Figure 12 shows the pressure trends according to the wheel angular position monitored upstream ($p_{IN,CFM}$) and downstream ($p_{OUT,CFM}$) of the mass flow meter. As can be seen from the pressure trends, the CFM is affected by the pump operation even if it is positioned upstream with respect to the gear pump. Pressure pulsations generate pressure forces that affect the CFM body, as reported in Figure 12. The magnitude of the force (F_{CFM}) is lower than 3 N but is variable according to the angular position of the wheel. This effect could generate some inaccuracies and disturbances during the mass flow rate measuring process.

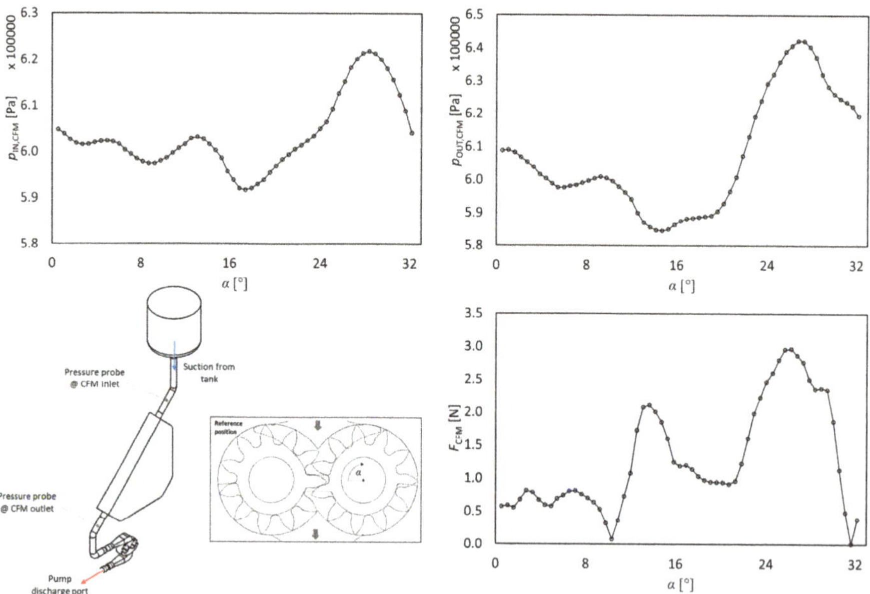

Figure 12. Pressure pulsation at the inlet and outlet sections of the CFM and the force trends according to the angular position of the gear pump wheel (see reference position).

In addition to the magnitude, the frequency of this pressure pulsation could also affect the performance of the CFM. Figure 13 shows the Fast Fourier Transformation (FFT) of the force signal reported in Figure 12. Each harmonic is reported according to its frequency, which corresponds to a multiple of the first harmonic obtained by considering the number of teeth (11) and the rotational velocity (450 rpm) adopted in the present analysis.

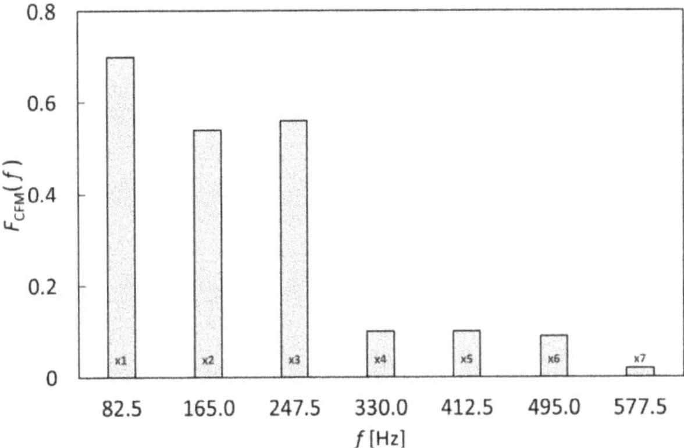

Figure 13. Harmonics magnitude and frequencies obtained by FFT of the force signal.

The magnitude of each harmonic is mainly based on the interaction between the wheels, relief groove located in the meshing region and the cavitation phenomenon. The FFT analysis permits the detection of possible dangerous interactions between the gear pump and the CFM operation. The presence of the cavitation phenomenon could generate a specific harmonic that may influence the CFM measurement performance. As reported in Figure 13, the third harmonic (corresponding to 247.5 Hz) is higher than the second one (corresponding to 165.0 Hz), representing a different excitation regime with respect to the natural one, which is easily determined by knowing the rotational speed and the number of teeth. This means that for certain operating conditions (depending on the rotational velocity and saturation pressure) the frequency of the flow disturbance may affect the CFM operation in terms of accuracy and stability [13,14]. Even if modern CFM adopts the high oscillation frequency of the measuring tubes to ensure that the correct operation of the measuring system is not influenced by pipe vibrations, analysis of the installation issues is preferable. In fact, the flow disturbance, as well as other noise (e.g., variable drive), could upset the precise measurement of a Coriolis sensor. Therefore, even with the aforementioned limitations involved in the present approach, the dynamic simulation allows for the analysis of the interaction between a single component (e.g., gear pump) and a measurement device (e.g., Coriolis flow meter). From the CFD results, it is possible to highlight several guidelines for reducing the aforementioned drawbacks:

- check the resonance frequency of the CFM installed against the pump operating range, including off-design conditions;
- cavitation phenomena could change the amplitude of the harmonics, determining different disturbance on the CFM operation according to the ORC operating point. For this reason, a proper design of the gear pump capacity could reduce the cavitation issues. Higher pump capacity allows for a reduction in the rotational velocity, limiting the cavitation issues and the corresponding high-frequency phenomenon (explained in detail in the following section);
- looking at the measurement performance, a stable pump operation can be assumed as a sort of offset with respect to the ideal measurement condition. When fluid dynamic effects (such as cavitation) occur, the measurement performance could fall down, reducing the chances to correct the measurement after the calibration process due to the non-linear interaction between the gear pump and the CFM;
- the use of check valves could reduce the backflow, generating a more stable operating condition of the gear pump.

4.2. Cavitation and Its Effects on the Control System and ORC Operating Point

The possibility of simulating the interaction of several components within an ORC cycle allows for the assessment of the control system performance and strategy, in particular: (i) characteristic time, (ii) priority and (iii) devices to be preserved and/or monitored for improving the reliability of the system. In fact, any control system is characterized by its own characteristic time, which is usually given by the component (active device or sensors), which shows the longest characteristic time coupled with the calculation and post-processing time of the data coming from the field. The decisions and actions taken by the control system must also meet some priority criteria (activation and/or control of devices) and satisfy certain criteria (boundaries) beyond which it is not possible to operate the system. In light of these considerations, the capability to simulate the interaction between two or more components represents valid support in this sense. A demonstration of such a powerful contribution given by the CFD calculation of the entire ORC system has already been reported in the literature [22,23], and a similar strategy is applied in this work for analyzing the gear pump operation.

Figure 14 shows an example of a control strategy that usually reacts to a sudden increase in the energy exchanged by the refrigerant fluid in the evaporator. This situation is quite common in ORC cycles, which are usually used to recover energy from energy waste, which is subject to unpredictable variations.

Following the diagram of Figure 14, starting from a steady-state operating condition, due to the greater energy exchanged in the evaporator, the refrigerant reaches a higher temperature upstream and downstream of the expander and, consequently, at the inlet of the condenser, the fluid will be warmer than the previous stationary condition. At the condenser, the thermal inertia of the system does not allow for an instantaneous reaction, generating an outlet liquid phase refrigerant with a lower degree of sub-cooling. For the liquid phase, a higher condenser outlet temperature implies higher saturation pressure and, consequently, greater cavitation effects.

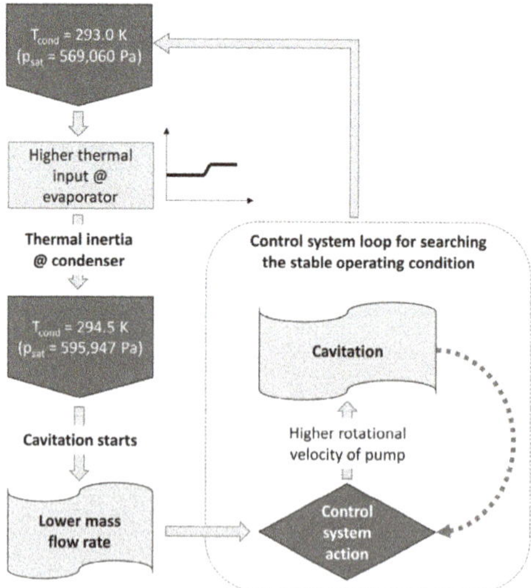

Figure 14. Logic diagram of the control system considering the local fluid dynamic behavior.

The phenomenon of cavitation (as described above) is closely related to the rotation speed of the pump, which is the parameter that the control system changes to adjust the flow, which is inevitably decreased as a result of the greater cavitation phenomenon. The control system, therefore, requires a greater rotation speed to restore the required flow rate searching the new operating point, for which a greater condenser cooling capacity will be required. Table 1 collects the conditions generated by the control system actions, and for which the numerical simulations are carried out. Pressure boundary conditions are the same of the previous analysis (corresponding to the operating point A of Table 1), only the liquid temperature, saturation pressure, and gear pump rotational speed are varied according to the previous description.

Figure 15 shows the contour plots of the gas fraction inside the gear pump, corresponding to the mid-plane of the tooth width for three different operating conditions: (i) the stationary operating points where the previous analyses are carried out (liquid temperature in the tank of 293.0 K, which corresponds to a saturation pressure of 569,000 Pa) and a rotational velocity of 450 rpm, (ii) the operating condition due to the instantaneous increment of the outlet temperature at the condenser (liquid temperature in the tank of 294.5 K, which corresponds to a saturation pressure of 596,000 Pa and a rotational velocity of 450 rpm, and, finally, (iii) the new operating conditions adjusted by the control system, with a rotating velocity equal to 475 rpm for a liquid temperature of 294.5 K. The quantification of the gas mass fraction allows for the evaluation of the contemporary effects of wheel motion (the rotational speed of the pump) and thermodynamic conditions (the saturation pressure of the tank). As described above, the contemporary effects of wheel rotation and temperature drop to generate a greater cavitation effect. From the control system point of view, the gear pump rotation is easy to measure and control in the opposite way of local temperature (or pressure), a result of the thermodynamic condition of the entire ORC cycle. For this reason, the analysis deals with the pump rotational speed as a driver of the cavitation phenomena reported in Figure 15. The similarities between the contour plots of Figure 15b,c are due to the value of the saturation pressure (and, thus, the temperature of the refrigerant in the tank), even if local phenomena (e.g., in the engagement region) are more intense in the worst case, characterized by the highest rotational speed (475 rpm).

Table 1. Operating conditions of the ORC system.

Operating Point	Liquid Temperature [K]	Saturation Pressure [Pa]	Pump Rotational Speed [rpm]
A	293.0	569,060	450
B	294.5	595,947	450
C	294.5	595,947	475

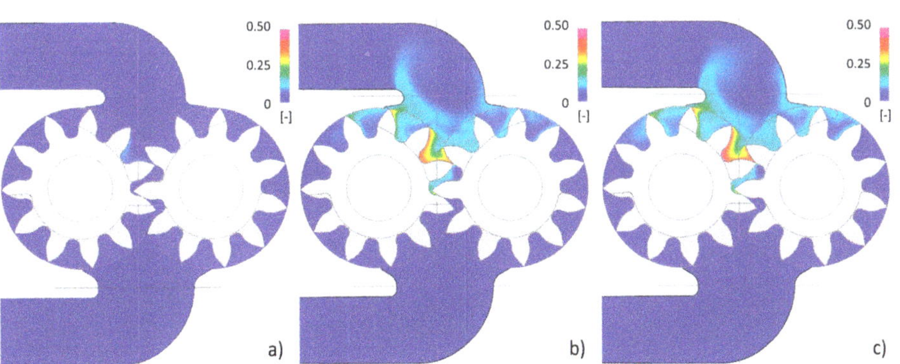

Figure 15. Contour plot of the gas mass fraction for a mid-plane of the tooth width according to three operating points: (**a**) condition A (450 rpm, 293.0 K), (**b**) condition B (450 rpm, 294.5 K) and (**c**) condition C (475 rpm, 294.5 K).

From the three plots shown in Figure 15, it is clear that the analysis of the interaction between the components of an ORC circuit influences its operation and the actions of the control system. The cavitation shown in Figure 15 is closely related to both the thermodynamic conditions and the piping layout of the present ORC system. The cavitation phenomenon is determined by the wheel rotation and the pressure losses that the liquid refrigerant experiences when passing through the

CMF and the suction port of the pump. For the stationary condition of Figure 15a, the gas fraction is reduced and confined only to the area downstream of the engaging region, while, when the sub-cooling decreases (see Figure 15b), the gas fraction involves a greater region at the pump suction port. From this condition, the mass flow rate decreases due to the reduced pump capacity that is affected by a certain amount of gas fraction at the suction port. The mass flow rate, processed by the pump according to the angular position of the gear wheel, is shown in Figure 16. As the operating conditions change, the flow rate processed changes, both in terms of the average value and the trend. In the stationary condition, the pump processes 0.091 kg/s. This average value decreases at the second operating point due to more severe cavitation conditions than previously. The average value of the mass flow rate, in this case, is equal to 0.082 kg/s (which corresponds to a 10% lower value than previously). From this condition, the control system increases the pump rotational speed to 475 rpm, bringing the average flow rate value back to 0.091 kg/s.

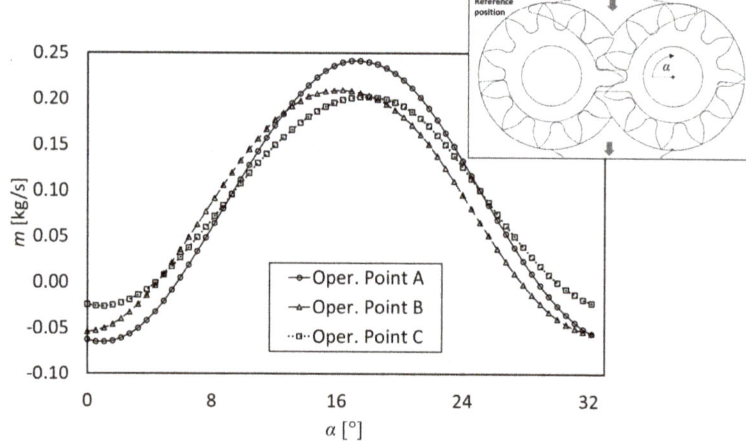

Figure 16. Mass flow rate trends according to the angular position of the wheel (see reference position) as a function of the three different operating conditions.

From the trends shown in Figure 16, it is clear that the new operating condition, even if characterized by an average flow rate identical to the starting regime condition, shows a different trend according to the angular position of the wheel. The presence of a greater gas fraction reduces the pulsation thanks to the greater compressibility of the liquid-gas mixture processed by the pump. Despite this smoother operating condition, cavitation could generate rapid erosion in some parts of the machine, especially at the axial gaps close to the suction port. In this region, the gas bubbles implode due to the localized increase in pressure.

5. Conclusions

In this paper, transient numerical simulations of a micro-ORC system branch have been carried out using a novel Whole Engine Model strategy. The aim of the present work was to analyze the fluid dynamic behavior of the gear pump operation on the system performance and control/measuring devices. The use of the dynamic model coupled with the dynamic mesh strategy allowed for the analysis of dynamic behavior such as pressure pulsation and cavitation phenomenon, related to the rotation and the engagement of gear pump wheels. With reference to the numerical results reported in this work, the main outcomes can be listed as follows:

- the operation of the gear pump in a micro-ORC system determines pressure pulsations that could affect the measurement systems and system performance;

- pressure pulsations affect the operation of the Coriolis flow meter installed downstream of the liquid receiver and upstream of the gear pump. The pressure pulsation results in an excitation force that is variable according to the gear pump wheel position;
- the sub-cooling degree affects the gear pump operation. The pressure losses generated by the piping, fittings and suction port geometry, together with the dynamic effects of wheel rotation, determine a pressure drop responsible for the cavitation phenomenon;
- the cavitation phenomenon is promoted by the decreasing sub-cooling degree (which determines the saturation pressure of the liquid phase) and by the increasing pump rotational speed.

From the present analysis, it is clear that the actions of the control system affect the operation of every single component of an ORC system. The analyses have shown that certain transient conditions could determine several risks for the reliability and performance of components. Measuring performance can be preserved/improved by following specific guidelines, which involve both the design criteria of the devices and the control strategy of the ORC system. For example, the instantaneous variation of the evaporator thermal input could determine a cavitation phenomenon in the pump, which is subject to the action of the control system that requests more flow rate for increasing the energy exchanged at the evaporator. In this sense, numerical simulations of an actual installation may represent a valid support for better control and management of the systems, improving the performance, increasing the operating life, and reducing the risks of system damage.

Author Contributions: A.S. conceptualization and writing original draft; E.F. and S.R. data curation; M.P. methodology and supervision; N.C. writing—review & editing.

Funding: This research was partially supported by the Italian Ministry of Economic Development within the framework of the Program Agreement MSE-CNR "Mi-cro co/tri generazione di Bioenergia Efficiente e Stabile (Mi-Best)".

Conflicts of Interest: The authors declare no conflict of interest.

Nomenclature

F	Force
m	mass flow rate
p	pressure
ω	pump rotational speed

Subscripts

cond	condenser
IN	inlet section
OUT	outlet section
sat	saturation (referred to)

Acronyms

CFD	computational fluid dynamics
CFM	Coriolis flow meter
FFT	Fast Fourier Transformation
ORC	organic Rankine cycle
WOM	whole ORC model

References

1. Landelle, A.; Tauveron, N.; Haberschill, P.; Revellin, R.; Colasson, S. Organic Rankine cycle design and performance comparison based on experimental database. *Appl. Energy* **2017**, *204*, 1172–1187. [CrossRef]
2. Macchi, E.; Astolfi, M. *Organic Rankine Cycle (ORC) Power Systems*, 1st ed.; Woodhead Publishing: Sawston, UK, 2016.
3. Tocci, L.; Pal, T.; Pesmazoglou, I.; Franchetti, B. Small scale Organic Rankine Cycle (ORC): A techno-economic review. *Energies* **2017**, *10*, 413. [CrossRef]

4. Bao, J.; Zhao, L. A review of working fluid and expander selections for organic Rankine cycle. *Renew. Sustain. Energy Rev.* **2013**, *24*, 325–342.

5. Wang, Y.; Chen, L.; Jia, B.; Roskilly, A.P. Experimental study of the operation characteristics of an air-driven free-piston linear expander. *Appl. Energy* **2017**, *195*, 93–99. [CrossRef]

6. Hou, X.; Zhang, H.; Xu, Y.; Yu, F.; Zhao, T.; Tian, Y.; Yang, Y.; Zhao, R. External load resistance effect on the free piston expander-linear generator for organic Rankine cycle waste heat recovery system. *Appl. Energy* **2018**, *212*, 1252–1261. [CrossRef]

7. Burugupally, S.P.; Weiss, L.; Depcik, C. The effect of working fluid properties on the performance of a miniature free piston expander for waste heat harvesting. *Appl. Therm. Eng.* **2019**, *151*, 431–438. [CrossRef]

8. Bianchi, M.; Branchini, L.; De Pascale, A.; Melino, F.; Ottaviano, S.; Peretto, A.; Torricelli, N. Application and comparison of semi-empirical models for performance prediction of a kW-size reciprocating piston expander. *Appl. Energy* **2019**, *249*, 143–156. [CrossRef]

9. Ichikawa, T.; Yamaguchi, K. On pulsation of delivery pressure of gear pump (in the case of a long delivery pipeline). *Bull. JSME* **1971**, *14*, 1304–1312. [CrossRef]

10. Kojima, E.; Shirada, M. Characteristic of fluidborne noise generated by fluid power pump (3rd report, discharge pressure pulsation of external gear pump). *Bull. JSME* **1984**, *27*, 2188–2195. [CrossRef]

11. Mucchi, E.; Dalpiaz, G.; Rivola, A. Dynamic behavior of gear pumps: Effect of variations in operational and design parameters. *Meccanica* **2011**, *46*, 1191–1212. [CrossRef]

12. Wang, T.; Baker, R. Coriolis flowmeters: A review of developments over the past 20 years, and an assessment of the state of the art and likely future directions. *Flow Meas. Instrum.* **2014**, *40*, 99–123. [CrossRef]

13. Vetter, G.; Notzon, S. Effect of pulsating flow on Coriolis mass flowmeters. *Flow Meas. Instrum.* **1994**, *5*, 263–273. [CrossRef]

14. Cheesewright, R.; Clark, C.; Hou, Y.Y. The response of Coriolis flowmeters to pulsating flows. *Flow Meas. Instrum.* **2004**, *15*, 59–67. [CrossRef]

15. Del Campo, D.; Castilla, R.; Raush, G.A.; Gamez Montero, P.J.; Codina, E. Numerical analysis of external gear pumps including cavitation. *J. Fluids Eng.* **2012**, *134*, 081105. [CrossRef]

16. Battarra, M.; Mucchi, E. Incipient cavitation detection in external gear pumps by means of vibro-acoustic measurements. *Measurement* **2018**, *129*, 51–61. [CrossRef]

17. Frosina, E.; Senatore, A.; Rigosi, M. Study of a high-pressure external gear pump with a computational fluid dynamic modeling approach. *Energies* **2017**, *10*, 1113. [CrossRef]

18. Yoon, Y.; Park, B.-H.; Shim, J.; Han, Y.-O.; Hong, B.-J.; Yun, S.-H. Numerical simulation of three-dimensional external gear pump using immersed solid method. *Appl. Therm. Eng.* **2017**, *118*, 539–550. [CrossRef]

19. Kolasiński, P.; Blasiak, P.; Rak, J. Experimental and numerical analyses on the rotary vane expander operating conditions in a micro organic Rankine cycle system. *Energies* **2016**, *9*, 606. [CrossRef]

20. Morini, M.; Pavan, C.; Pinelli, M.; Romito, E.; Suman, A. Analysis of a scroll machine for micro ORC applications by means of a RE/CFD methodology. *Appl. Therm. Eng.* **2015**, *80*, 132–140. [CrossRef]

21. Bianchi, M.; Branchini, L.; Casari, N.; De Pascale, A.; Melino, F.; Ottaviano, S.; Pinelli, M.; Spina, P.R.; Suman, A. Experimental analysis of a micro-ORC driven by piston expander for low-grade heat recovery. *Appl. Therm. Eng.* **2019**, *148*, 1278–1291. [CrossRef]

22. Casari, N.; Suman, A.; Ziviani, D.; Morini, M.; Pinelli, M. Virtual Model for ORC—Whole ORC Modeling: WOM. In Proceedings of the ECOS 2017 30th International Conference on Efficiency, Cost, Optimisation, Simulation and Environmental Impact of Energy Systems, San Diego, CA, USA, 2–6 July 2017.

23. Randi, S.; Casari, N.; Pinelli, M.; Suman, A.; Ziviani, D. WOM: Whole ORC Model. In Proceedings of the 17th International Refrigeration and Air Conditioning Conference at Purdue, Lafayette, IN, USA, 9–12 July 2018.

24. Simerics Inc. *PumpLinx's User Manual—V 4.6.4*; Simerics Inc.: Bellevue, WA, USA, 2018.

Article

Structured Mesh Generation and Numerical Analysis of a Scroll Expander in an Open-Source Environment

Ettore Fadiga *, Nicola Casari, Alessio Suman and Michele Pinelli

Department of Engineering (DE), University of Ferrara, 44122 Ferrara, Italy; nicola.casari@unife.it (N.C.); alessio.suman@unife.it (A.S.); michele.pinelli@unife.it (M.P.)
* Correspondence: ettore.fadiga@unife.it; Tel.: +39-0532-974964

Received: 31 December 2019; Accepted: 29 January 2020; Published: 4 February 2020

Abstract: The spread of the organic rankine cycle applications has driven researchers and companies to focus on the improvement of their performance. In small to medium-sized plants, the expander is the component that has typically attracted the most attention. One of the most used types of machine in this scenario is the scroll. Among the other methods, numerical analyses have been increasingly exploited for the investigation of the machine's behaviour. Nonetheless, there are major challenges for the successful application of computational fluid dynamics (CFD) to scrolls. Specifically, the dynamic mesh treatment required to capture the movement of working chambers and the nature of the expanding fluids require special care. In this work, a mesh generator for scroll machines is presented. Given few inputs, the software described provides the mesh and the nodal positions required for the evolution of the motion in a predefined mesh motion approach. The mesh generator is developed ad hoc for the coupling with the open-source CFD suite OpenFOAM. A full analysis is then carried out on a reverse-engineered commercial machine, including the refrigerant properties calculations via CoolProp. It is demonstrated that the proposed methodology allows for a fast simulation and achieves a good agreement with respect to former analyses.

Keywords: scroll; opensource CFD; OpenFOAM; CoolFOAM; WOM; positive displacement machine; expander; ORC

1. Introduction

Scroll compressors have been extensively employed in air conditioning and refrigeration since the 1980s. Their success is mostly related to a low level of noise and vibrations, together with a small number of moving parts and a compact design. In the last decade, scroll compressors and other positive displacement machines (PDMs, e.g., piston, screw, and vane machines) have been operated as expanders in Organic Rankine Cycles (ORCs) to generate power from waste heat and renewable energies [1]. Compared to its alternatives, the scroll expander is generally characterized by higher pressure ratios and efficiency and by lower flow rates and rotational speeds [2]. On the other hand, one of the major drawbacks of this technology is the maximum working temperature (maximum temperature of 250 °C reported by Seher et al. [3]): higher temperatures would increase excessively the thermal expansion of scroll spirals, leading to significant increments of internal leakages [4].

Scroll expanders are frequently adopted in micro-ORCs with power outputs up to 2 kW, as described in different literature works reporting experimental tests. Wang et al. [5] have tested a scroll expander with R134a over a wide range of rotational speeds, reaching a shaft work output close to 1 kW. Experimental characterizations of scroll expanders with R245fa are described in References [6,7], with maximum isentropic efficiencies of 66.5% and 75.7%, respectively. Other tests using HCFC-123 and R134a as working fluids are analyzed in References [8,9].

The architecture of the scroll geometry is the cornerstone of the design and optimization of the whole machine. The inventor of this technology, Léon Creux, has introduced in 1905 a scroll profile based on the involute of a circle, which is still the most spread solution [10]. However, various alternative profile shapes have been developed during the XX century [11–13].

The expander is designed with a fixed built-in volume ratio, which is the volume of the discharge chamber at the beginning of the discharge phase divided by the volume of the suction chamber at its maximum extension. The built-in volume ratio determines the theoretical pressure ratio of the fluid expansion. Nonetheless, leakage flows, off-design operating conditions, and wear could commonly force the expander to work in under- or over-expansion conditions. The expander works in conditions of under-expansion when the internal pressure ratio imposed by the chambers volumes variation is lower than the system pressure ratio. Consequently, the pressure in the working chamber at the end of the expansion is higher than the pressure in the discharge area. This phenomenon produces a significant flow rate peak at the expander output when the discharge phase occurs. On the contrary, over-expansion occurs when the internal pressure ratio of the expander is higher than the system pressure ratio. The fluid is then forced to flow back, leading to a recompression, which is particularly detrimental for the machine performance [8].

Currently, the most used methods for the design and the prediction of scroll expanders performance are based on theoretical modeling, thermodynamic analysis, and experimental studies. Shaffer et al. [14] have presented a control volume approach to model the geometry of a scroll machine. The analytical and thermodynamic modeling of scroll compressors is extensively treated in References [15–19], with and without experimental validation. For what concerns the scroll expander, Ma et al. [20] have presented a dynamic model with experimental validation, introducing an overall dynamic friction coefficient of the machine to enhance the model adaptability. Lately, Bell et al. and Ziviani et al. have presented [21] and demonstrated the capabilities [22] of an open-source general framework for the simulation of various positive displacement machines, including scrolls. When the geometry of the scroll expander is not exactly known, deterministic models similar to the ones reported above are not suitable options. Lemort et al. [8] have developed and validated a semi-empirical model that requires a limited number of parameters. Examples of similiar studies for scroll and other positive displacement expanders are reported in References [23–27].

Recently, industries and researches have began to include Computational Fluid Dynamics (CFD) analyses in scroll machine designs and optimizations [28–33]. This tool allows to retrieve information about leakage flows nature, temperature distribution, and three-dimensional behavior of the flow inside the expander/compressor. Moreover, CFD could be employed to tune analytical and thermodynamic models as low-cost alternatives to experimental campaigns. One of the most challenging aspects of computational analyses of scroll machines (and positive displacement machines in general) is the grid generation process.

The nature of the problem imposes to adopt a dynamic mesh approach in order to correctly model the moving parts of the machines. Casari et al. [34] have reviewed different approaches for the dynamic mesh treatment in screw machines numerical analyses, including immersed boundary, adaptive remeshing, and key-frame remeshing. Most of the techniques presented in their work are suitable also for scroll expanders and compressors, but the most popular approach is certainly the custom predefined mesh generation [35]. This technique consists in the realization of a user-defined number of grids per rotor pitch. Each grid represents a set of control points (coordinates) through which the mesh nodes has to pass during the simulation. Several examples of this approach can be found in the literature, including applications to screw machines [36–39], root machines [40,41], and scroll machines [28,31,42,43].

In this work, the authors present a C++ library for the generation of structured body-fitted meshes of scroll compressors and expanders. Starting from the scroll profiles and main dimensions, the meshes are generated by means of an elliptic grid generation algorithm. This library has been realized in accordance with the coding standards of the open-source CFD software OpenFOAM. Furthermore, the authors have performed a full three-dimensional CFD analysis of a scroll expander suitable for micro-ORC applications. The library developed can be a useful tool for the development of structured meshes of scroll machines. One of the possible outcomes of this work could be the realization of advanced numerical analysis of ORC systems, such as the Whole ORC Model (WOM) presented in Reference [44].

2. Materials and Methods

The authors have realized a C++ library according to the OpenFOAM coding standard in order to perform numerical simulations of scroll machines by means of the custom predefined mesh generation method. The first step in this method is the realization of a body-fitted structured grid starting from the scroll geometry. This kind of grid is characterized by elements of good quality and well-defined connectivity rules. When the path of the moving parts is known, a set of structured meshes can be generated in advance in order to describe chronologically sequential positions of the moving bodies.

In this case, an increment in time corresponds to an increment of the crank angle, so the user is able to define a number of angular positions of the mobile spiral that have to be meshed. When the actual simulation is executed, the nodes of these predefined meshes become control points for the grid nodes in order to maintain good element quality. The mesh nodes pass through the control points, but the connectivity of the mesh remains the same. In this way, both re-meshing and interpolations are avoided during the analysis. One of the advantages of this strategy is that it is respectful of the space conservation law. Moreover, mass and energy imbalances derived from interpolation procedures are not a concern in this case because no interpolation is performed.

2.1. Body-Fitted Structured Grid

The structured grid generation process of a scroll machine has been achieved as follows:
STRUCTURED DOMAIN DEFINITION

The generation of a structured mesh of all the scroll domains, including the ports region, is a very challenging task. It could be very difficult to model narrow gaps and complex features of the geometry. Consequently, it has been decided to isolate a portion of the machine that includes the mobile scroll, as represented in Figure 1.

Figure 1. Structured domain of the scroll machine.

The two edges of the structured domain are imported as a list of points, representing the control points for two different spline curves. This region has been discretized by means of a structured strategy, while the other parts can be meshed with unstructured algorithms. Then, the connection between the two regions must be performed with one of the interface strategies implemented in OpenFOAM.

MOBILE SPIRAL DISCRETIZATION

The mesh boundary discretization starts from the mobile spiral edge, where the boundary points are defined according to criteria based on the variation of the slope of the straight line tangent to the spiral. The variation of the slope is maintained constant along the mobile spiral in order to have a finer discretization in the central region of the machine, where the high pressure port is placed.

The position of the points on the mobile edge remains the same during the whole meshing process. The points are translated with the spiral in order to represent different angular position, but their relative positions do not change.

FIXED SPIRAL AND CASING DISCRETIZATION

The other edge of the structured domain is represented by the fixed spiral merged with a portion of the casing profile. The points are collocated on the fixed edge in order to create lines as normal to the edges as possible.

In the central region of the machine, where it is not possible to create lines normal to both the edges, a special treatment is introduced: the points on the fixed edge follow a squared cosine distribution with variable parameters (related to phase and period) in function of the crank angle. The user has to find the best fitting parameters for few main angular positions of the mobile scroll, and then, the intermediate parameters are interpolated from the main ones. The first guess for the internal nodes is then provided by a transfinite interpolation.

ELLIPTIC INTERNAL MESH GENERATION

The second step is an elliptic mesh generator with a control map that imposes the orthogonality of the grid at the boundary. This method has been developed according to the work presented by Spekreijse [45]. This meshing strategy is characterized by a transformation from a computational space (ξ,η) to the cartesian domain (x,y), passing through a parameter space (s,t), as represented in Figure 2.

The Picard iteration method with a finite difference discretization has been applied for the solution of a system of nonlinear elliptic equations, reported below:

$$P^{k-1}x_{\xi\xi}^{k} + 2Q^{k-1}x_{\xi\eta}^{k} + R^{k-1}x_{\eta\eta}^{k} + S^{k-1}x_{\xi}^{k} + T^{k-1}x_{\eta}^{k} = 0$$

where x represents the position vector and the subscripts ξ and η represent a derivative in function of the variables of the computational space. The apexes k and k-1 refer to the current and previous iteration of the Picard method, respectively. In this system of equations, the coefficients P, Q, and R can be calculated using the derivatives of the position vectors at the previous iteration in the function of the computational space variables. On the contrary, in order to evaluate S and T, the derivative of the position vectors relative to the parameter space variables are needed. The variables of the parameter space are calculated by means of the following Laplace equations:

$$\Delta s(\xi,\eta) = 0$$

$$\Delta t(\xi,\eta) = 0$$

where s and t represent the unknowns of the systems while the orthogonality at the boundary is reached by imposing a Neumann boundary condition at the boundaries of interest. The Laplace equations have been discretized by means of the finite volume strategy, and the resulting systems have been solved using a BiConjugate Gradient Stabilized (BiCGSTAB) method.

The structured mesh is generated in two dimensions and then extruded in order to create a three-dimensional computational grid.

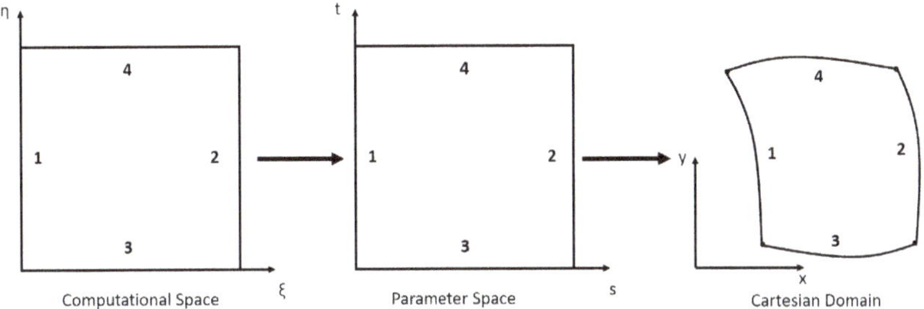

Figure 2. Transformation from computational space to cartesian domain. Adapted from Reference [45].

2.2. Moving the Mesh

The first structured mesh is generated for an initial position of the scroll machine. Later, the user has to define a number of grids per revolution that the software must generate and to write in files containing the list of control points for the mesh nodes. A dynamic motion solver, which has been added to the set available in OpenFOAM, updates the nodal positions for any user-defined time step, keeping the control points files as a reference. In order to evaluate the location of a mesh node for the ith time step, the following equation is solved:

$$x_{i,final} = \alpha * x_{i,AG} + (1 - \alpha) * x_{i,NG}$$

where the subscript AG represents the actual grid file and NG is the next grid file. AG and NG are the files including the nodal positions at the time steps we are interpolating among.

The library that has been developed in this work is therefore comprehensive of two steps: mesh generation (including the actual mesh and the nodal position for a user-defined number of angular steps) and the mesh motion part, that takes care of the evolution of the geometry as the operation of the machine proceeds. The sum of the two goes under the name of the *ScrollFOAM* library and is the only addition to be provided to OpenFOAM in order to simulate scroll machines (either in compression or expansion configuration) with the above-described approach. The library, as developed, can be fully integrated into OpenFOAM and can be empolyed with factory or user-defined libraries (e.g., for the CoolFOAM wrapper, as described below).

2.3. Scroll Geometry and Simulation Setup

At this stage, the mesh has been completed (both the starting position as well as the control points that drive the evolution), and therefore, the simulation can be set up. The procedure is shown on an expander that is set to operate in realistic conditions. The cycle taken as a reference is a μ-ORC installed at the University of Bologna and is described in Reference [46]. The original cycle is composed of a brazed plate heat exchanger with 64 plates as evaporator, a brazed plate heat exchanger with 16 plates as a regenerator, a shell and tube heat exchanger as condenser, a volumetric three pistons radial engine used as expander, and a volumetric gear pump controlled by an inverter that supplies the organic fluid over the ORC system position under the liquid receiver (tank) realizing a column of water of about 1 m high. The ORC system operates with R134a as working fluid. This layout is considered as a reference and is exploited to feed the numerical model with actual operating conditions.

The scroll expander simulated in this work is thought to be a potential replacement for the piston engine. The design point for the expander has been chosen for the analysis, as it falls inside the operating range of the cycle. Specifically, the machine investigated is the commercial SANDEN TRSA09-3658 scroll compressor. The geometry of the model has been obtained by means of a reverse engineering procedure, as described in Reference [47]. The acquired model is then slightly modified: the thickness of the spirals has been reduced of 1/7th with respect to the original work of Reference [47].

This change has been introduced for the sake of computational speed: all the features of the scroll are retained, and the simulation is fully 3D. Also, the position of the discharge port is modified: the port has been redesinged to be axial as the inlet and to be located in proximity to the discharge. This expedient has been included in order to avoid the presence of a non-conformal interface to link the morphing chamber domain to the static part constituted by the casing. The non-conformal interface approach has been tested with unsatisfactory results: in that scenario, the interface would unavoidably deform in the normal-to-interface direction. This occurrence is not robustly handled by OpenFOAM, leading to stability problems. The turbulence model has found to be severely affected by this problem. The resulting fluid domain is reported in Figure 3.

Figure 3. Fluid domain of the scroll expander comprehensive of the high pressure (central) and discharge ports.

The built-in volume ratio of the machine is of 1.82, given by two spirals of 2 wraps each. A single working chamber per spiral moves the refrigerant from the high pressure port to the low one. The geometry tested does not present axial clearances. The flank gaps that characterize the operation of the machine are reported in Table 1. The size of the clearances is fully in line with common engineering practice.

Table 1. Flank gap size evolution during the operation of the machine.

Position	Gap Size (μm)
0° (inlet chambers close)	20
90°	36
180°	94
270°	36

The boundary conditions that have been imposed at the inlet of the scroll are static temperature equal to 358 K, far field static pressure of 20 bar (wave transmissive), 1% of turbulent intensity, and a mixing length of 0.7 mm (the inlet channel diameter is equal to 12 mm). A static pressure of 11 bar fixes the counter pressure at which the machine operates. All the walls are considered adiabatic. The simulation runs fully turbulent with the k-ω SST model of Menter [48] with updated coefficients [49].

Concerning the thermophysical property calculations of the refrigerant during the operation of the expander, both the ideal gas model with constant properties and the full real gas calculation have been tested. The refrigerant operates in a region in which the deflection from the ideality is nonnegligible. For this reason, the equation of state proposed by Tillner-Roth et al. has been adopted [50]. The modeling is carried out thanks to the CoolFOAM wrapper developed by the authors [51]: the CoolProp library [52] is exploited for retrieving the properties in the actual conditions. This choice has been characterized by an increase of 23% in the elapsed computational time of one revolution, if compared to the ideal gas approximation.

3. Results

3.1. Mesh Results

The final result is a mesh with satisfying values of skewness, orthogonality, and aspect ratio. All these parameters, and many others, widely fit all the quality criteria set by the OpenFOAM standards. An example of structured mesh is presented in Figure 4, which represents the discretization of a SANDEN TRSA09-3658 scroll compressor. Table 2 contains some illustrative values of skewness, non-orthogonality, and aspect ratio for this grid. More details on this machine, which has been numerically analyzed in this work, are given in the next sections of this paper.

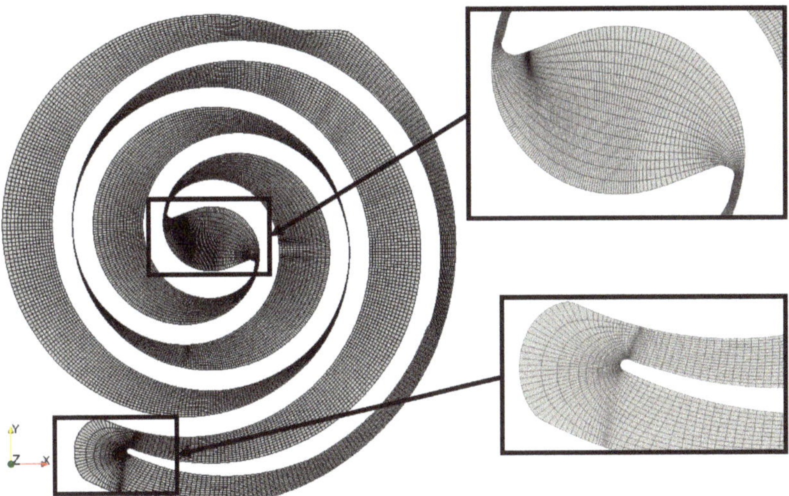

Figure 4. Structured mesh: final result.

Table 2. Mesh quality parameters in function of the mobile scroll position.

Position	Max. Skewness	Max. Non-Orthogonality	Avg. Non-Orthogonality	Max. Aspect Ratio
0°	2.28	78.23	8.04	863.25
90°	2.39	60.96	7.41	528.44
180°	2.45	50.30	7.61	219.34
270°	2.58	50.99	7.90	200.98

3.2. CFD Results

The pressure pattern as the operation of the scroll evolves with the conditions described in Section 2.3 is reported in Figure 5. The filling procedure of the high-pressure chamber can be noticed passing from Figure 5a–d. Further, the good operation in proximity of the design point is proven by the pressure of the chamber before the opening of the discharge port (Figure 5d shows the pressure distribution of a few degrees before the actual opening): the pressure in the chamber is slightly above the output value, and therefore, no under or overexpansion is recorded. The minimum pressure reached in the domain is slightly above 8 bar and reached downstream the working chamber sealing clearance. In such areas, the flow becomes chocked as reported in Reference [34]. The simulation catches well such phenomena and proves to be robust and stable also in case of transonic problems.

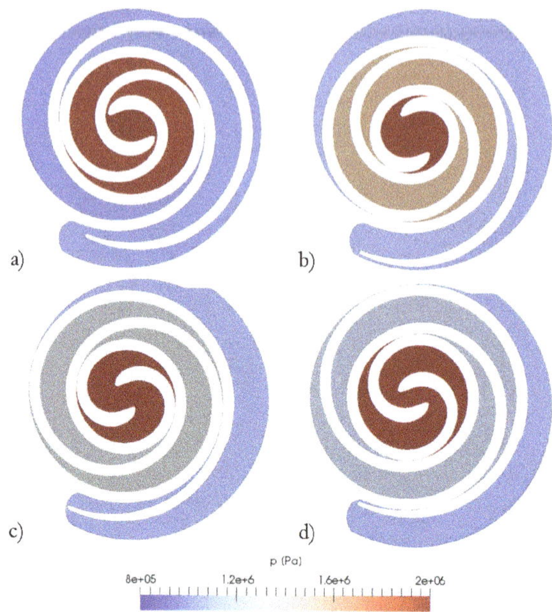

Figure 5. Pressure pattern evolution during the operation of the expander: (**a**) 0°, (**b**) 90°, (**c**) 180°, and (**d**) 270°.

Another interesting result shows the deflection from the ideal behaviour of the refrigerant during the operation and is reported in Figure 6. The maximum deflection from 1 is found in the inlet chamber and is roughly 15%. As a general remark, the real gas effects lower (Z tends to 1) as the fluid goes through the gaps and increase in the bulk of the chambers.

Figure 6. Compressibility factor for R134a during the operation of the expander.

Regarding the performance of the scroll, Figure 7 reports the mass flow rate that is processed over an entire revolution of the machine. Under the conditions investigated, the machine processes 27.5 g/s. It should be kept in mind that only a portion of the entire thickness is included. The value obtained is in line with other investigation of the current machine [47]. It is possible to estimate also the VFM (Volumetric Flow Matching Ratio), as the theoretical volume is isolated by the scroll over the CFD-calculated volume flow operated by the scroll. The resulting VFM is of 1.23. This value is perfectly in line with typical values of VFM, that ranges from 1.07 to 1.3 [53]. The imbalance in the mass with the current setup is of the order of 0.1%, that is remarkably good considering the application and the presence of two ACMIs (Arbitrary Coupling Mesh Interfaces). The outflow shows high pulsations with respect to the inlet, and the amplitude of the oscillation is of roughly 80% of the average value.

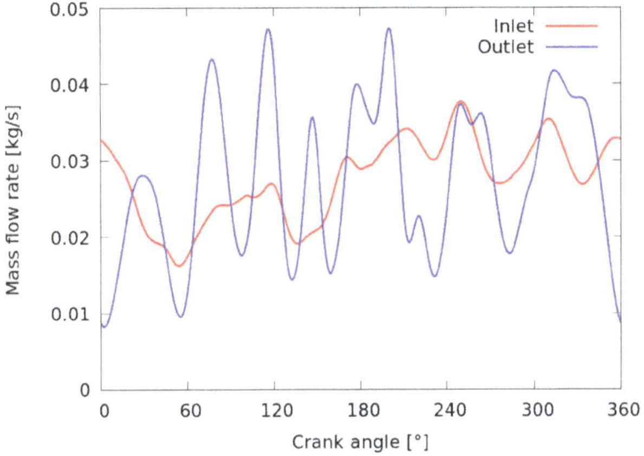

Figure 7. Flow rate variation over a revolution of the shaft.

With regards to the output that can be extracted from the machine, as reported in Figure 8, the power varies over a period from less than 50 W to almost 250 W, with an average of 149.2 W. This value has been calculated as the product between the mechanical torque and the rotational speed (2000 rpm). This is perfectly in line with what has been found in the previous simulation and with the nameplate power of the machine (that is 1 kW, very close to 7 times the actual power). The average internal isentropic efficiency of the machine is therefore 41%, that is typical for this kind of applications [53]. This value has been calculated as the ratio of the average power produced by the expander and the power that it would produce if the expansion of the fluid was isentropic.

A final remark regards the impact of the gaps in the volumetric efficiency. To help in visualizing the relative importance of the gaps, Figure 9 reports the flow field in a constant-span plane. It can be clearly seen how the velocity in clearances is remarkably high with respect to the bulk of the working chambers. An interesting remark regards the different speeds that are reached in the two gaps of each pocket: the higher Mach numbers are obtained in the second gap, closer to the outlet port. From Figure 9, it can also be seen that the high speed traces downstream the gaps are the ones in which the real gas effects are lower (as can be noticed by comparison with Figure 6).

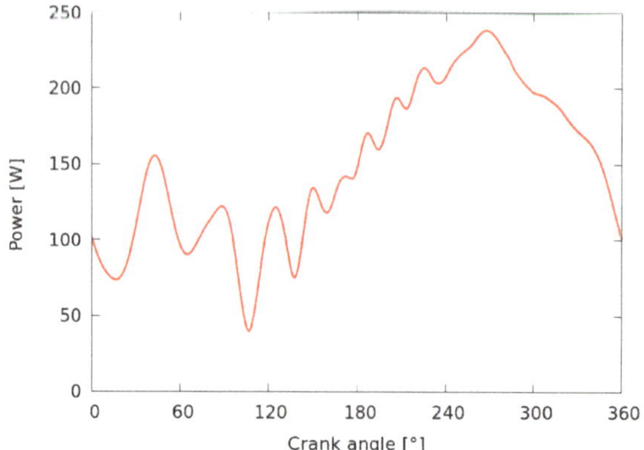

Figure 8. Power variation over a revolution of the shaft.

Figure 9. Velocity distribution in the domain: The arrow size is scaled by the local velocity magnitude.

4. Conclusions

In this work, a new meshing tool developed in C++ has been proposed. The capability of this tool to manage a fully 3D CFD simulation is proven by the analysis of a commercial scroll compressor in an open-source CFD environment, i.e., OpenFOAM. The results that are presented here are in line with what has been found in previous numerical investigations of the same scroll machine. Those calculations, on the other hand, have been performed with commercial software, and this work provides a tool for performing CFD analysis in a positive displacement machine with open-source software. Moreover, an imbalance in mass of the order of 0.1%, which represents a very good level of accuracy, has been reached.

The application of *ScrollFOAM* provides a unique tool for the numerical investigation of refrigerant behaviour in scroll machine with open-source tools. The procedure has been demonstrated for an expander. This tool can be used to foresee the off-design behaviour of the ORC cycle in terms of output power performance. Variations in the boundary conditions have been tested, and the solver proves to be robust and accurate in catching over- and under-expansions. The suitability of the entire infrastructure to the investigation of ORC cycles is enhanced by the coupling of the solver with CoolFOAM, the CoolProp wrapper for OpenFOAM. With such an expedient, the real gas modeling gains in accuracy as the Helmholtz equation of state can be used. The tool developed in this work is a fundamental step for achieving the WOM infrastructure, as the state-of-the-art in fast and robust simulation of positive displacement machines in OpenFOAM is limited to piston-type expanders. The extension of the modeling capabilities to other types of machines is needed for a more spread usability of CFD in ORC cycles.

Author Contributions: E.F. and N.C. methodology, conceptualization, and writing original draft; E.F. software and data curation; M.P. supervision and funding acquisition; A.S. writing—review and editing. All authors have read and agreed to the published version of the manuscript.

Funding: The research was partially supported by the Italian Ministry of Economic Development within the framework of the Program Agreement MSE-CNR "Micro co/tri generazione di Bioenergia Efficiente e Stabile (Mi-Best)".

Conflicts of Interest: The authors declare no conflict of interest.

Abbreviations and Nomenclature

Symbols

x	nodal position
α	blending factor

Subscript and Superscript

i	*i*th time step
final	final position of the grid node

Acronyms

AG	Actual Grid
CFD	Computational Fluid Dynamics
NG	Next Grid
ORC	Organic Rankine Cycle
PDM	Positive Displacement Machine
WOM	Whole ORC Model

References

1. Landelle, A.; Tauveron, N.; Haberschill, P.; Revellin, R.; Colasson, S. Organic Rankine cycle design and performance comparison based on experimental database. *Appl. Energy* **2017**, *204*, 1172–1187, [CrossRef]
2. Emhardt, S.; Tian, G.; Chew, J. A review of scroll expander geometries and their performance. *Appl. Ther. Eng.* **2018**, *141*, 1020–1034, [CrossRef]
3. Seher, D.; Lengenfelder, T.; Gerhardt, J.; Eisenmenger, N.; Hackner, M.; Krinn, I. Waste heat recovery for commercial vehicles with a Rankine process. In Proceedings of the 21st Aachen Colloquium on Automobile and Engine Technology, Aachen, Germany, 8–10 October 2012; pp. 7–9.
4. Lemort, V.; Legros, A. 12—Positive displacement expanders for Organic Rankine Cycle systems. In *Organic Rankine Cycle (ORC) Power Systems*; Macchi, E., Astolfi, M., Eds.; Woodhead Publishing: Sawston, UK, 2017; pp. 361–396, [CrossRef]
5. Wang, H.; Peterson, R.B.; Herron, T. Experimental performance of a compliant scroll expander for an organic Rankine cycle. *Proc. Inst. Mech. Eng. Part A J. Power Energy* **2009**, *223*, 863–872, [CrossRef]
6. Álvarez-Alvarado, J.M.; Ríos-Moreno, G.J.; Ventura-Ramos, E.; Ronquillo-Lomelí, G.; Trejo-Perea, M. Experimental Study of a 1-kW Organic Rankine Cycle Using R245fa Working Fluid and a Scroll Expander: A Case Study. *IEEE Access* **2019**, *7*, 154515–154523, [CrossRef]
7. Declaye, S.; Quoilin, S.; Guillaume, L.; Lemort, V. Experimental study on an open-drive scroll expander integrated into an ORC (Organic Rankine Cycle) system with R245fa as working fluid. *Energy* **2013**, *55*, 173–183, [CrossRef]
8. Lemort, V.; Quoilin, S.; Cuevas, C.; Lebrun, J. Testing and modeling a scroll expander integrated into an Organic Rankine Cycle. *Appl. Ther. Eng.* **2009**, *29*, 3094–3102, [CrossRef]
9. Ali Tarique, M.; Dincer, I.; Zamfirescu, C. Experimental investigation of a scroll expander for an organic Rankine cycle. *Int. J. Energy Res.* **2014**, *38*, 1825–1834, [CrossRef]
10. Creux, L. Rotary Engine. U.S. Patent No. 801,182, 15 April 1905.
11. Guttinger, H. Displacement machine for Compressible Media. U.S. Patent No. 3,989,422, 29 January 1976.

12. Young, N.; McCullough, J. Scroll Type Positive Fluid Displacement Apparatus. U.S. Patent No. 3,884,599, 20 May 1975.

13. Montelius, C. Rotary Compressor or Motor. US Patent No. 2,324,168, 13 July 1943.

14. Shaffer, B.R.; Groll, E.A. Variable wall thickness scroll geometry modeling with use of a control volume approach. *Int. J. Refrig.* **2013**, *36*, 1809–1820, [CrossRef]

15. Peng, B.; Lemort, V.; Legros, A.; Hongsheng, Z.; Haifeng, G. Variable thickness scroll compressor performance analysis—Part I: Geometric and thermodynamic modeling. *Proc. Inst. Mech. Eng. Part E J. Process Mech. Eng.* **2017**, *231*, 633–640, [CrossRef]

16. Peng, B.; Zhao, S.; Li, Y. Thermodynamic Model and Experimental Study of Oil-free Scroll Compressor. *J. Phys. Conf. Ser.* **2017**, *916*, 012048, [CrossRef]

17. Chen, Y.; Halm, N.P.; Groll, E.A.; Braun, J.E. Mathematical modeling of scroll compressors—part I: compression process modeling. *Int. J. Refrig.* **2002**, *25*, 731–750, [CrossRef]

18. Bell, I.H.; Groll, E.A.; Braun, J.E.; Horton, W.T.; Lemort, V. Comprehensive analytic solutions for the geometry of symmetric constant-wall-thickness scroll machines. *Int. J. Refrig.* **2014**, *45*, 223–242, [CrossRef]

19. Bell, I.H. Theoretical and Experimental Analysis of Liquid Flooded Compression in Scroll Compressors. Ph.D. Thesis, Purdue University, West Lafayette, IN, USA, 2011.

20. Ma, Z.; Bao, H.; Roskilly, A.P. Dynamic modelling and experimental validation of scroll expander for small scale power generation system. *Appl. Energy* **2017**, *186*, 262–281, [CrossRef]

21. Bell, I.H.; Ziviani, D.; Lemort, V.; Bradshaw, C.R.; Mathison, M.; Horton, W.T.; Braun, J.E.; Groll, E.A. PDSim: A general quasi-steady modeling approach for positive displacement compressors and expanders. *Int. J. Refrig.* **2020**, *110*, 310–322, [CrossRef]

22. Ziviani, D.; Bell, I.H.; Zhang, X.; Lemort, V.; Paepe, M.D.; Braun, J.E.; Groll, E.A. PDSim: Demonstrating the capabilities of an open-source simulation framework for positive displacement compressors and expanders. *Int. J. Refrig.* **2020**, *110*, 323–339, [CrossRef]

23. Mendoza, L.; Lemofouet, S.; Schiffmann, J. Testing and modelling of a novel oil-free co-rotating scroll machine with water injection. *Appl. Energy* **2017**, *185*, 201–213, [CrossRef]

24. Bianchi, M.; Branchini, L.; De Pascale, A.; Melino, F.; Ottaviano, S.; Peretto, A.; Torricelli, N. Performance prediction of a reciprocating piston expander with semi-empirical models. *Energy Procedia* **2019**, *158*, 1737–1743, [CrossRef]

25. Giuffrida, A. Improving the semi-empirical modelling of a single-screw expander for small organic Rankine cycles. *Appl. Energy* **2017**, *193*, 356–368, [CrossRef]

26. Bianchi, M.; Branchini, L.; De Pascale, A.; Melino, F.; Ottaviano, S.; Peretto, A.; Torricelli, N. Application and comparison of semi-empirical models for performance prediction of a kW-size reciprocating piston expander. *Appl. Energy* **2019**, *249*, 143–156, [CrossRef]

27. Fanelli, E.; Pinto, G.; Cornacchia, G.; Braccio, G. Parameters identification for scroll expander semi-empirical model by using genetic algorithm. *Energy Procedia* **2018**, *148*, 736–743, [CrossRef]

28. Gao, H.; Ding, H.; Jiang, Y. 3D Transient CFD Simulation of Scroll Compressors with the Tip Seal. *IOP Conf. Ser. Mater. Sci. Eng.* **2015**, *90*, 012034, [CrossRef]

29. Wang, J.; Song, Y.; Li, Q.; Zhang, D. Novel structured dynamic mesh generation for CFD analysis of scroll compressors. *Proc. Inst. Mech. Eng. Part A J. Power Energy* **2015**, *229*, 1007–1018, [CrossRef]

30. Suman, A.; Randi, S.; Casari, N.; Pinelli, M.; Nespoli, L. Experimental and Numerical Characterization of an Oil-Free Scroll Expander. *Energy Procedia* **2017**, *129*, 403–410, [CrossRef]

31. Sun, S.; Wu, K.; Guo, P.; Yan, J. Analysis of the three-dimensional transient flow in a scroll refrigeration compressor. *Appl. Ther. Eng.* **2017**, *127*, 1086–1094, [CrossRef]

32. Song, P.; Zhuge, W.; Zhang, Y.; Zhang, L.; Duan, H. Unsteady Leakage Flow Through Axial Clearance of an ORC Scroll Expander. *Energy Procedia* **2017**, *129*, 355–362, [CrossRef]

33. Picavet, A.; Genevois, D. Three-Dimensional Navier-Stokes Simulations of Working of Scroll Compressors. In Proceedings of the 24th International Refrigeration and Air Conditioning Conference at Purdue, West Lafayette, IN, USA, 9–12 July 2018.

34. Casari, N.; Suman, A.; Ziviani, D.; Van Den Broek, M.; De Paepe, M.; Pinelli, M. Computational Models for the Analysis of positive displacement machines: Real Gas and Dynamic Mesh. *Energy Procedia* **2017**, *129*, 411–418. [CrossRef]

35. Rane, S. Grid Generation and CFD Analysis of Variable Geometry Screw Machines. Ph.D. Thesis, City University London, London, UK, 2015.

36. Rane, S.; Kovacevic, A. Algebraic generation of single domain computational grid for twin screw machines. Part I. Implementation. *Adv. Eng. Softw.* **2017**, *107*, 38–50. [CrossRef]

37. Kovacevic, A.; Stosic, N.; Smith, I. *Screw Compressors: Three Dimensional Computational Fluid Dynamics and Solid Fluid Interaction*; Springer Science & Business Media: Berlin, Germany, 2007; Volume 46.

38. Kovacevic, A.; Rane, S. Algebraic generation of single domain computational grid for twin screw machines Part II–Validation. *Adv. Eng. Softw.* **2017**, *109*, 31–43. [CrossRef]

39. Basha, N.; Rane, S.; Kovacevic, A. Multiphase Flow Analysis in an Oil-injected Twin Screw Compressor. In Proceedings of the 3rd World Congress on Momentum, Heat and Mass Transfer (MHMT'18), Budapest, Hungary, 12–14 April 2018 .

40. Casari, N.; Fadiga, E.; Pinelli, M.; Suman, A.; Kovacevic, A.; Rane, S.; Ziviani, D. Numerical investigation of oil injection in a Roots blower operated as expander. *IOP Conf. Ser. Mater. Sci. Eng.* **2019**, *604*, 012075, [CrossRef]

41. Singh, G.; Sun, S.; Kovacevic, A.; Li, Q.; Bruecker, C. Transient flow analysis in a Roots blower: Experimental and numerical investigations. *Mech. Syst. Signal Process.* **2019**, *134*, 106305, [CrossRef]

42. Ding, H.; Jiang, Y. CFD Simulation of An Oil Flooded Scroll Compressor Using VOF Approach. In Proceedings of the 23th International Refrigeration and Air Conditioning Conference at Purdue, West Lafayette, IN, USA, 11–14 July 2016.

43. Hesse, J.; Spille-Kohoff, A.; Andres, R.; Hetze, F. CFD simulation of scroll compressors with axial and radial clearances and thermal deformation. In *18 Internationales Stuttgarter Symposium*; Bargende, M., Reuss, H.C., Wiedemann, J., Eds.; Springer Fachmedien Wiesbaden: Wiesbaden, Germany, 2018; pp. 123–137.

44. Randi, S.; Casari, N.; Pinelli, M.; Suman, A.; Ziviani, D. WOM: Whole ORC Model. In Proceedings of the 24th International Refrigeration and Air Conditioning Conference at Purdue, West Lafayette, IN, USA, 9–12 July 2018.

45. Spekreijse, S. Elliptic Grid Generation Based on Laplace Equations and Algebraic Transformations. *J. Comput. Phys.* **1995**, *118*, 38–61, [CrossRef]

46. Bianchi, M.; Branchini, L.; Casari, N.; De Pascale, A.; Melino, F.; Ottaviano, S.; Pinelli, M.; Spina, P.; Suman, A. Experimental analysis of a micro-ORC driven by piston expander for low-grade heat recovery. *Appl. Ther. Eng.* **2019**, *148*, 1278–1291. [CrossRef]

47. Morini, M.; Pavan, C.; Pinelli, M.; Romito, E.; Suman, A. Analysis of a scroll machine for micro ORC applications by means of a RE/CFD methodology. *Appl. Ther. Eng.* **2015**, *80*, 132–140. [CrossRef]

48. Menter, F.R.; Esch, T. Elements of Industrial Heat Transfer Prediction. In Proceedings of the 16th Brazilian Congress of Mechanical Engineering (COBEM), Uberlândia, Brazil, 26–30 November 2001.

49. Menter, F.R.; Kuntz, M.; Langtry, R. Ten years of industrial experience with the SST turbulence model. *Turbul. Heat Mass Transf.* **2003**, *4*, 625–632.

50. Tillner-Roth, R.; Baehr, H.D. An International Standard Formulation for the Thermodynamic Properties of 1,1,1,2-Tetrafluoroethane (HFC-134a) for Temperatures from 170 K to 455 K and Pressures up to 70 MPa. *J. Phys. Chem. Ref. Data* **1994**, *23*, 657–729, [CrossRef]

51. Fadiga, E.; Casari, N.; Suman, A.; Pinelli, M. CoolFOAM: The CoolProp wrapper for OpenFOAM. *Comput. Phys. Commun.* **2019**, 107047. [CrossRef]

52. Bell, I.H.; Wronski, J.; Quoilin, S.; Lemort, V. Pure and pseudo-pure fluid thermophysical property evaluation and the open-source thermophysical property library CoolProp. *Ind. Eng. Chem. Res.* **2014**, *53*, 2498–2508. [CrossRef] [PubMed]

53. Mathias, J.A.; Johnston, J.R.; Cao, J.; Priedeman, D.K.; Christensen, R.N. Experimental testing of gerotor and scroll expanders used in, and energetic and exergetic modeling of, an organic Rankine cycle. *J. Energy Resour. Technol.* **2009**, *131*, 012201. [CrossRef]

Article

Regression Models for the Evaluation of the Techno-Economic Potential of Organic Rankine Cycle-Based Waste Heat Recovery Systems on Board Ships Using Low Sulfur Fuels [†]

Enrico Baldasso [1,*], Maria E. Mondejar [1], Ulrik Larsen [2] and Fredrik Haglind [1]

[1] Department of Mechanical Engineering, Technical University of Denmark, 2800 Kgs. Lyngby, Denmark; maemmo@mek.dtu.dk (M.E.M.); frh@mek.dtu.dk (F.H.)

[2] Department of Administration, Copenhagen University, 1165 Copenhagen, Denmark; ula@adm.ku.dk

[*] Correspondence: enbald@mek.dtu.dk; Tel.: +45-45-25-41-05

[†] This is an extended version of our paper published in: Baldasso E., Mondejar M.E., Larsen U., and Haglind F. Prediction of the annual performance of marine organic Rankine cycle power systems. In Proceedings of the 31st International Conference on Efficiency, Cost, Optimization, Simulation and Environmental Impact of Energy Systems, Guimarães, Portugal, 17–22 June 2018.

Received: 28 February 2020; Accepted: 14 March 2020; Published: 16 March 2020

Abstract: When considering waste heat recovery systems for marine applications, which are estimated to be suitable to reduce the carbon dioxide emissions up to 20%, the use of organic Rankine cycle power systems has been proven to lead to higher savings compared to the traditional steam Rankine cycle. However, current methods to estimate the techno-economic feasibility of such a system are complex, computationally expensive and require significant specialized knowledge. This is the first article that presents a simplified method to carry out feasibility analyses for the implementation of organic Rankine cycle waste heat recovery units on board vessels using low-sulfur fuels. The method consists of a set of regression curves derived from a synthetic dataset obtained by evaluating the performance of organic Rankine cycle systems over a wide range of design and operating conditions. The accuracy of the proposed method is validated by comparing its estimations with the ones attained using thermodynamic models. The results of the validation procedure indicate that the proposed approach is capable of predicting the organic Rankine cycle annual energy production and levelized cost of electricity with an average accuracy within 4.5% and 2.5%, respectively. In addition, the results suggest that units optimized to minimize the levelized cost of electricity are designed for lower engine loads, compared to units optimized to maximize the overall energy production. The reliability and low computational time that characterize the proposed method, make it suitable to be used in the context of complex optimizations of the whole ship's machinery system.

Keywords: organic Rankine cycle; low sulfur fuels; waste heat recovery; regression model; predictive model; ship; techno-economic feasibility; machinery system optimization

---

1. Introduction

The increasing awareness of the environmental impact of the shipping industry is pushing the development of novel solutions to reduce the emission of pollutants from ships. In this context, the International Maritime Organization (IMO) recently introduced a novel legislation framework constraining the emissions of nitrogen oxides (NO_x) [1] and sulfur oxides (SO_x) [2], and set the vision to reduce the greenhouse gas (GHG) emissions by 50% compared to the emissions levels of 2008 by 2050 [3].

Several alternative solutions can, in theory, lead to a reduction of the emissions from shipping, one of them being the use of liquefied natural gas (LNG) [4]. Compared to the use of traditional heavy fuel oils (HFO), the use of LNG leads to substantial reductions in the NO_x and SO_x emissions [5]: the former can be reduced up to 85% thanks to the lean combustion process, while the latter are almost completely eliminated, because LNG does not contain sulfur. Additionally, a significant potential to reduce the environmental impact of ships lies in the possibility to recover the waste heat released by the ship's engine system. According to the work from Bouman et al. [6], the implementation of waste heat recovery (WHR) solutions on board a vessel can result in a reduction of the carbon dioxide (CO_2) emissions by up to 20%.

Waste-to-heat recovery solutions are particularly attractive for cruise ships [7,8], because such ships require a significant amount of heat for internal uses, while waste-to-power solutions are the most commonly investigated for containerships and tankers, among others [9].

The organic Rankine cycle (ORC) power system has been identified as a promising solution for waste-to-power recovery applications on board vessels [9], because it enables the attainment of higher design power outputs [10] and improved off-design efficiencies [11], in comparison with the traditional steam Rankine cycle technology.

Most of the previous works dealing with the optimal design of ORC power cycles for maritime applications have been following a model-driven approach, meaning that they rely on the use of numerical models based on the laws of thermodynamics, and validated by comparison with experimental data, when available. Such models are then used to investigate the performance of the unit under different design and off-design conditions.

The development of suitable numerical models to predict the performance and optimal design of ORC units for maritime applications is, however, a challenging task, because the designer needs to account for a multitude of aspects, including the availability of multiple waste heat sources, the ORC off-design performance, the ship sailing profile, the impact of the WHR unit on the engine's performance, and the presence of dynamic instabilities, among others.

In this regard, Soffiato et al. [12] described a procedure to integrate the use of multiple waste heat sources on board a LNG-carrier. Baldi et al. [13] discussed multiple optimization approaches and concluded that the ship sailing profile needs to be accounted for during the ORC design procedure in order to maximize the annual energy production. Michos et al. [14] highlighted the impact of the additional backpressure supplied to the exhaust line of a marine engine by the implementation of a WHR unit, while Baldasso et al. [15] suggested that constraining the maximum backpressure on the exhaust side of the ORC WHR boiler leads to a significant reduction of the attainable power output from the recovery unit. Lastly, Rech et al. [16] pointed out the importance of carrying out dynamic simulations to ensure that the system is correctly sized and is capable of reaching steady-state operation under a wide range of engine operating conditions.

The downsides of the approach followed by the aforementioned works lie in the model complexity, and in the computational time required to carry out the simulations, which increases as more aspects are taken into account. These approaches are therefore commonly used in a limited number of case studies, but are not suitable to carry out preliminary estimations of the potential for installing ORC-based WHR units on board a wide range of vessels.

To cover this need, another range of studies are available in the literature. Other authors, in fact, aimed at deriving simplified methodologies to estimate the performance of ORC units. Among these works, Liu et al. [17] proposed an equation to estimate the thermal efficiency of an ORC unit based on the working fluid evaporation, condensation, and critical temperatures. Kuo et al. [18] correlated the Jakob number with the attainable ORC thermal efficiency, while Wang et al. [19] used the Jakob number in predictive models to estimate the ORC thermal and exergy efficiencies. Larsen et al. [20] proposed the use of multiple regression models to predict the ORC thermal efficiency given the boundary conditions of the process. Lecompte et al. [21] focused on the cycle second law efficiency and derived a simplified correlation to estimate its maximum value for a given waste heat source.

Lastly, Palagi et al. [22] developed surrogate models based on a neural network approach to carry out multi-objective optimization of a small-scale ORC unit, showing that the computational time could be reduced by two orders of magnitude in comparison with the traditional optimization approach.

These works pose a solid foundation for the rapid estimation of the prospects for installing an ORC unit, but have two limitations: (1) they aim at estimating the unit efficiency, rather than its power output, which is the key performance indicator in case of WHR applications, and (2) they are suitable for design conditions estimations only, thus, they do not consider the off-design performance of the unit, which is of key importance when considering the maritime application, because the waste heat availability changes according to the engine load.

To the best of the authors' knowledge, only one previously published work describes a simplified method to estimate the off-design performance of an ORC unit: Dickes et al. [23] carried out experimental and numerical investigations on a 2 kW$_{el}$ ORC system featuring a scroll expander and proposed a set of equations to characterize the optimal off-design operation of the ORC system. As the equations were derived based on a specific unit, their general applicability is, however, not guaranteed. In addition, the off-design characterization of ORC units featuring volumetric expanders is not comparable to the one of units featuring turbo-expanders (i.e., units tailored for maritime applications).

This work aims at deriving a set of models for the quick and accurate prediction of the annual energy production and economic attractiveness of ORC units optimized for marine applications, considering low-sulfur fuels. The models were derived by numerical regression of a synthetic database of optimized ORC units and their effectiveness was tested in several test cases, where the estimations of the simplified approaches are compared with the solutions of the thermodynamic simulations.

The main novel contributions of the work are: (1) the derivation of off-design performance curves applicable to ORC units of different sizes and operating at different design conditions, and (2) a method to combine design and off-design performance curves for the ORC optimal design point that maximizes either the annual energy production, or the economic effectiveness estimated by means of the levelized-cost-of-electricity (LCOE).

The proposed method is not computationally intensive, and is therefore suitable to be used in the context of large optimization problems, such as ship routing optimizations, and holistic optimization and evaluation of a ship machinery system [24]. It is expected that the proposed method could support industry, decision makers, and researchers in the identification of the most feasible cases to integrate WHR units as part of the ship's machinery system.

The article is structured as follows: Section 2 describes the applied methods. The attained results are presented in Section 3 and discussed in Section 4. Finally, conclusions are outlined in Section 5.

2. Methods

The overall method to carry out simplified evaluations for the prospects for installing ORC units on board vessels powered by low-sulfur fuels was built by implementing the following steps: (1) a dataset of ORC design power and part-load performance was generated; (2) regression models were developed based on the calculated data; and (3) the reliability of the proposed regression models was tested in two case studies comparing the outputs of the thermodynamic models with outputs of the simplified regression curves. The following subsections detail the approach used for each step.

2.1. ORC Models

The evaluations were carried out considering a simple non-recuperated ORC unit recovering heat from the exhaust gases of the ship's main engine and using the seawater as heat sink. The power generated by the turbine is then converted into electricity in a generator. Figure 1 shows the sketch of the considered unit layout and the corresponding T-s diagram.

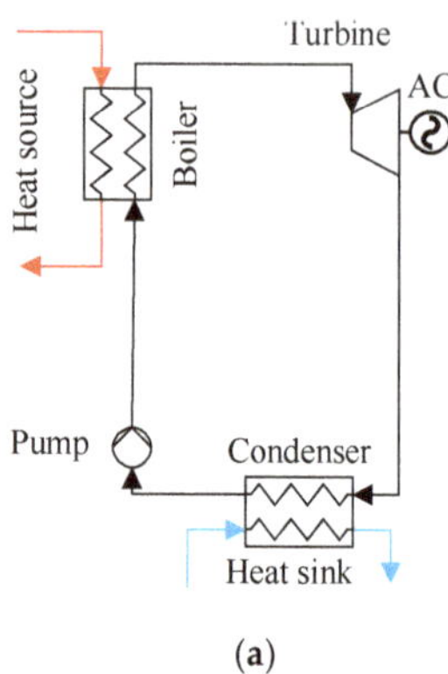
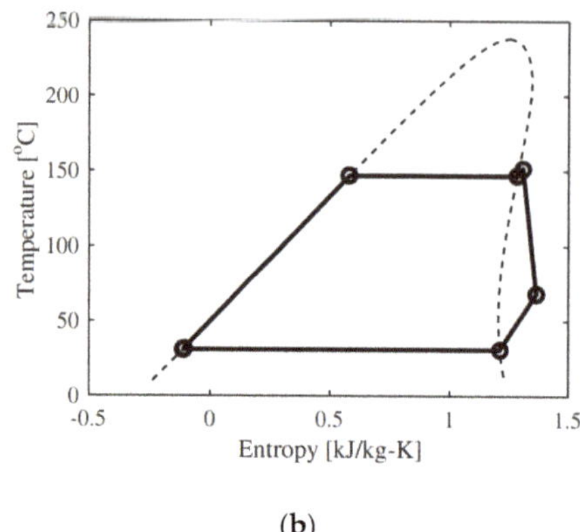

(**a**)   (**b**)

Figure 1. Considered organic Rankine cycle (ORC): (**a**) layout, (**b**) T-s diagram.

The estimation of the ORC design power output was carried out using the numerical model described in Andreasen et al. [25]. The model was previously validated by comparing its estimations with the results of other numerical studies published in the literature, indicating its suitability to estimate the ORC first and second law efficiencies with a maximum relative deviation of 3.3%.

The ORC net power output ($\dot{W}_{Net}$) was calculated as follows:

$$\dot{W}_{Net} = \dot{W}_{exp}\,\eta_{gear}\,\eta_{gen} - \dot{W}_p - \dot{W}_{p,sw}, \tag{1}$$

where η_{gear} and η_{gen} are the efficiencies of the gearbox and the electric generator, while the subscripts exp, p, and sw stand for expander, pump, and seawater. $\dot{W}_{p,sw}$ represents the power consumption of the pump supplying seawater to the condenser.

The ORC design model was used to identify the ORC cycle parameters maximizing the cycle net power output for every given heat source/sink combination. Table 1 lists the decision variables considered in the optimization procedure. A minimum superheating degree of 5 °C was imposed to ensure full evaporation of the working fluid, while the wide ranges were selected for the ORC mass flow rate, and condensation temperature were selected so that the optima always lies in between the boundaries. Because ships using low-sulfur fuels are considered, no design constraints were imposed to the ORC boiler in order to prevent issues related to possible sulfur condensation [9].

Table 1. ORC design model: decision variables in the optimization procedure.

Decision Variable	Lower Bound	Upper Bound
Turbine inlet pressure (bar)	1	$0.8\,P_{crit}$
ORC superheating (°C)	5	50
ORC mass flow rate (kg/s)	0.2	60
Condensation temperature (°C)	5	60

Additionally, the maximum and minimum allowed cycle pressures were set to 3000 and 4.5 kPa, according to the recommendations by Rayegan et al. [26], Drescher and Brüggeman [27], and MAN Energy Solutions [28]. Cyclopentane was selected as a working fluid in all cases, because it was previously shown to be a suitable working fluid candidate for maritime applications [11,15], and its properties were retrieved using Coolprop 4.2.5 [29]. Table 2 shows the parameters that were kept fixed during the optimization procedure. The properties of the exhaust gases were assumed to be equal to those of air at 100 kPa.

Table 2. ORC design model: fixed parameters.

Parameter	Value
Heat source inlet temperature (°C)	170–320
Heat source mass flow rate (kg/s)	5–120
Cooling water inlet temperature (°C)	5–30
Cooling water temperature increase (°C)	5
Minimum pinch point in the boiler (°C)	15–25
Minimum pinch point in the condenser (°C)	5–10
Turbine isentropic efficiency (-)	0.85
Pump isentropic efficiency (-)	0.7
Gearbox efficiency (-)	0.98
Electric generator efficiency (-)	0.98
Pressure drop in the heat exchangers (bar)	0
Pressure rise across sea water pump (bar)	2

All the optimizations were carried out using a combination of particle swarm (swarm size = 1000, number of generations = 50) and pattern search (500 iterations) available in the Matlab optimization toolbox [30].

The off-design performance of the ORC units was estimated by using the numerical model described in Baldasso et al. [15], whose estimations were validated in comparison with the experimental campaign carried out during the PilotORC project [31]. The comparison between the numerical estimations and the experimental values indicated that the model is suitable to predict the ORC power output, pressure levels, and mass flow rate with an accuracy within 5%.

The ORC units were assumed to be operated with a sliding pressure strategy during part-load operation, and the both superheating at the turbine inlet and the cooling water mass flow rate were kept constant.

2.2. Economic Evaluations

The economic performance of the ORC unit was evaluated by means of the levelized cost of electricity of the generated electricity, computed as [32]:

$$LCOE = \frac{I_0 + \sum_{y=1}^{n} \frac{O\&M_y}{(1+r)^y}}{\sum_{y=1}^{n} \frac{E_y}{(1+r)^y}}, \tag{2}$$

where the calculation considers that the system is operated for a number of years equal to $n = 25$. The symbols $O\&M_y$ and E_y represent the maintenance costs and the electricity generation at the year y. The symbol I_0 represents the initial investment cost. The electricity generation represents the annual energy production from the ORC unit. The annual $O\&M$ costs were set to 1.5% the ORC capital cost, while the discount rate (r) was set to 6%. The $O\&M$ costs were assumed to be the only operating costs of the system, as waste heat is used as heat input to the ORC unit.

Different estimation techniques are possible for the quantification of the investment cost related to the purchase of the ORC unit. The most common techniques are based on the estimation of the cost of the various components, which is dependent on their size, materials, and operating pressure [33]. These approaches require, however, a preliminary sizing of the components. Cost estimation can therefore represent a complex task. Here, we included a simplified costing approach based on the size of the ORC unit. The work from Lemmen [34] includes a list of real and estimated specific costs for the installation of ORC units whose size ranges from 200 kW to 8000 kW. Figure 2 reports the ORC-specific costs connected with the installation of ORC units for WHR applications, which were reported by Lemmens [34], and the regression curve which was used in this work.

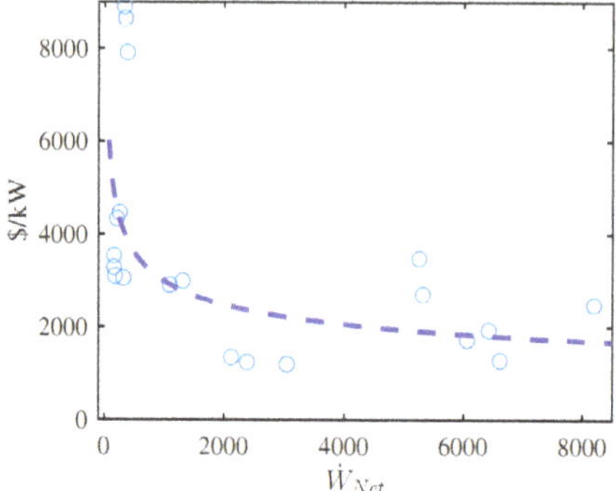

Figure 2. ORC-specific cost as a function of its size. The data was retrieved from Lemmens [34].

The costs presented in the previous work were available in €$_{2014}$ and were here converted into US dollar ($\$_{2015}$) using the Chemical Engineering Plant Cost Index (CEPCI) and a Euro-to-US dollar conversion factor of 1.1. The attained regression curve describing the variation of the ORC-specific cost as a function of its capacity is the following:

$$ORC_{specific\ cost}(\$) = 19,358 \cdot \dot{W}_{Net}{}^{-0.2703},$$

(3)

2.3. Data Generation and Regression Models

The regression surfaces for the estimation of the ORC design point power output were obtained by fitting the results of a dataset attained by running 200 independent ORC design optimizations, based on random design parameters. Two regression surfaces were attained:

- Design regression model #1, where the sampled parameters were the heat source inlet temperature and mass flow rate, as well as the cooling water inlet temperature.
- Design regression model #2, where also the boiler and condenser pinch point temperatures ($\Delta T_{pp,boil}$ and $\Delta T_{pp,cond}$) were included as sampled parameters in the optimization routines.

In both cases, the samples of the ORC design parameters were generated using the Sobol method [35] to ensure a good coverage of the sampling space, and the sampled parameters were constrained to be within the boundaries described in Table 2. Only the samples leading to ORC units with a power output in the range 250 kW to 2500 kW were considered in the regression procedure, and this ensured that the accuracy of the attained regression curves was not affected by the presence of outliers.

With respect to the boundaries defined for the heat source temperature, mass flow rate, and the cooling water temperature, a comparison with data retrieved from the MAN Energy Solutions CEAS calculation tool [36], suggests that the engine waste heat characteristics are within the considered ranges, when considering engines in the range from 5 MW to 50 MW operating with different tuning techniques and sailing both in cold and warm waters.

The off-design performance of every optimized ORC configuration was evaluated in 20 different and randomly generated off-design conditions. Each half of the randomly generated off-design points were imposed a temperature of the exhaust gases higher and lower than the design point, respectively. The exhaust gas mass flow rate was varied within 25% to 100% of the ORC design mass flow rate. The heat source temperature was allowed a deviation of +/- 80 °C compared to the design value. The sea water temperature was kept constant in all the off-design simulations.

A single regression model was attained for the characterization of the ORC off-design performance, considering an ORC load ranging for 10% to 100%, and based on the ORC designs used for both design models (fixed and variable pinch point temperatures).

Figure 3 provides a graphical overview of the procedure used to derive the regression models. The symbols included in the sketch are explained in Section 3.1 (T_{pp} refers both to the condenser and boiler pinch point temperature).

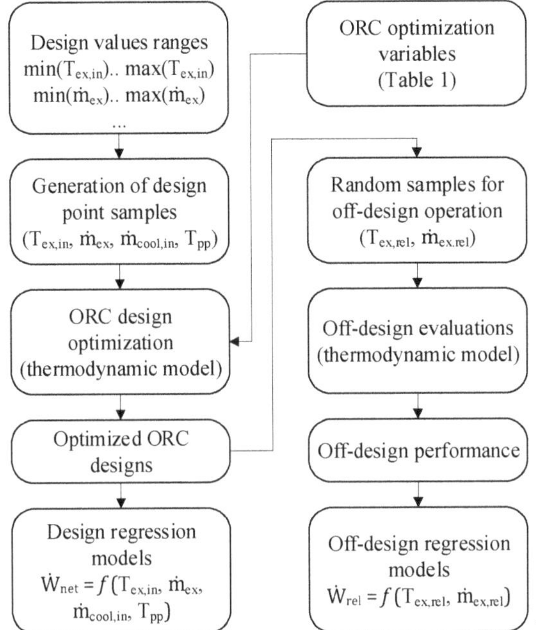

Figure 3. Procedure used to derive the ORC performance regression models.

The statistical robustness of a regression curve is ensured if the residuals follow a normal distribution and their mean is equal to zero. In addition, there should be no correlation between the residuals themselves and the parameters used to build the regression curve, nor the data points that are being estimated (the ORC net power output and its off-design performance). The validity of the mentioned aspects for the proposed regression surface models was checked with the scatter plots shown in the following sections.

2.4. Case Studies and Optimization Approach

The accuracy of the proposed regression surface models was checked in two test cases. The first case study considers the installation of an ORC unit on board an LNG-fueled feeder ship powered by a 10.5 MW MAN 7S60E-C10.5-GI engine with low pressure selective catalytic reactor tuning. In the second case study, an ORC unit on board a medium size LNG-fueled container vessel powered by a 23 MW MAN 6S80ME-C9.5-GI engine with part-load tuning was considered. The two vessels were assumed to operate according to the load profiles shown in Figure 4, for 4380 and 6500 hours annually, respectively.

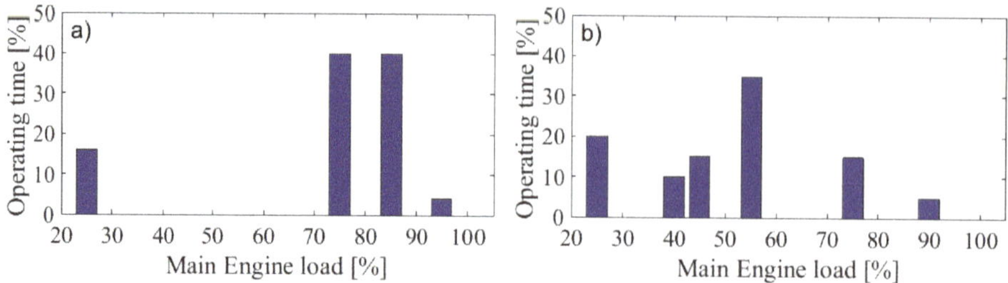

Figure 4. Considered annual engine load profiles: (**a**) Feeder, (**b**) Container vessel.

These are typical data for the two considered types of vessels. Sea water temperatures of 10 °C, 15 °C, and 20 °C were considered for the case studies.

Table 3 reports the exhaust mass flow rate and temperature for the two engines as a function of the load. The data was retrieved from the MAN CEAS engine calculation tool [36].

Table 3. Characteristics of the exhaust gases of the two engines as a function of the engine load [36].

	MAN 7S60E-C10.5-GI		MAN 6S80ME-C9.5-GI	
Engine Load (%)	**Mass Flow Rate (kg/s)**	**Temperature (°C)**	**Mass Flow Rate (kg/s)**	**Temperature (°C)**
100	22.5	251	52	251
90	19	266	47	239
80	18.2	245	44.9	207
70	16.3	243	40.3	206
60	14.3	248	35.7	211
50	12.1	258	30.6	221
40	9.9	271	25	239
30	7.5	280	22.7	205

The ORC maximum annual energy production and minimum LCOE required to produce the electricity by means of the ORC unit, were computed both using thermodynamic model and regression model. Both the thermodynamic calculations and the estimations using the regression models were carried out following the procedure shown in Figure 5.

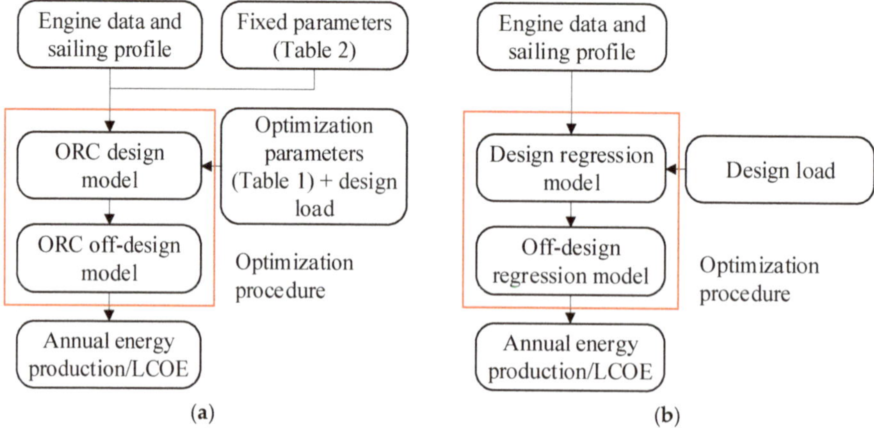

Figure 5. Procedure used to optimize the ORC annual energy production/LCOE: (**a**) thermodynamic models, (**b**) regression models.

When using the thermodynamic model, the ORC design parameters and design point were selected as optimization parameters, while only the latter was optimized when using the simplified approach based on the regression curves. For the estimation of the accuracy of the regression model #2, three alternative pinch point temperature combinations were considered: (i) $\Delta T_{pp,boil} = 25\ ^\circ C$ and $\Delta T_{pp,cond} = 10\ ^\circ C$, (ii) $\Delta T_{pp,boil} = 20\ ^\circ C$ and $\Delta T_{pp,cond} = 8\ ^\circ C$, and (iii) $\Delta T_{pp,boil} = 15\ ^\circ C$ and $\Delta T_{pp,cond} = 5\ ^\circ C$.

3. Results

This section presents the results obtained throughout the study. First, the set of regression equations is presented. Second, the results of the estimations of model #1 (fixed pinch point values) and thermodynamic models are compared. Third, the comparison focuses on model #2 (variable pinch point values) and the thermodynamic evaluations. Both comparisons analyze the ORC annual energy production and LCOE.

3.1. Regression Models

The fitted regression models to estimate the ORC design power output are given in Equation (4) and Equation (5), for model #1 and model #2, respectively. Note that the pinch point temperature differences are included as parameters in Equation (5). Equation (6) displays the regression model employed to fit the ORC off-design performance, i.e., extending both models #1 and #2.

$$\dot{W}_{Net} = a + b \cdot \frac{\dot{m}_{ex} \cdot T_{ex,in}}{1000} + c \cdot \dot{m}_{ex} \cdot \frac{\left(T_{ex,in} - T_{cool,in}\right)^3}{10,000,000} \tag{4}$$

$$\dot{W}_{Net} = a + b \cdot \frac{\dot{m}_{ex} \cdot T_{ex,in}}{1000} + c \cdot \dot{m}_{ex} \cdot \frac{\left(T_{ex,in} - T_{cool,in}\right)^3}{10,000,000} + d \cdot \frac{\dot{m}_{ex} \cdot T_{ex,in} \cdot \Delta T_{pp,boil}}{10,000} + e \cdot \frac{\dot{m}_{ex} \cdot T_{cool,in} \cdot \Delta T_{pp,cond}}{100}, \tag{5}$$

$$\dot{W}_{rel} = \frac{\dot{W}_{off}}{\dot{W}_{Net}} = a + b \cdot \sqrt{\dot{m}_{ex,rel}} + c \cdot \dot{m}_{ex,rel} \cdot T^2_{ex,rel}, \tag{6}$$

The off-design regression curve estimates the ORC relative net power output ($\dot{W}_{rel}$) in comparison with the design point conditions. The selected predictors are the relative exhaust gases mass flow rate ($\dot{m}_{ex,rel}$) and temperature ($T_{ex,rel}$), defined as follows:

$$\dot{m}_{ex,rel} = \frac{\dot{m}_{ex,off}}{\dot{m}_{ex,des}}, \tag{7}$$

$$T_{ex,rel} = \frac{T_{ex,in,off}}{T_{ex,in,des}}, \tag{8}$$

where the subscripts 'des' and 'off' refer to design and off-design conditions, respectively. Table 4 shows the regression coefficients and standard errors for the two proposed regression curves. The standard errors of each of the coefficients are smaller than the coefficient themselves, suggesting that all the coefficients were identified with high accuracy. The 'a' coefficients for design model #1 and #2 have relatively high standard error compared to the other parameters. The corresponding P-values are equal to 0.043 (model #1) and 0.045 (model #2). The P-values represent the result of the P-test aiming at understanding whether a regression parameter is significant to the prediction. Given that all the P-values are below 0.05, it can be concluded that all the selected parameters are highly significant (as should be expected).

Table 4. Regression coefficients and standard errors of the proposed regression models.

	Design Model #1		Design Model #2		Off-Design Model	
	Value	Standard Error	Value	Standard Error	Value	Standard Error
a	11.2332	5.5082	11.6575	5.7680	−0.1372	0.0021
b	10.0910	0.5773	39.6980	1.3410	0.1420	0.0035
c	19.2098	0.1247	18.3483	0.1411	1.0439	0.0022
d	-	-	−10.6565	0.4166	-	-
e	-	-	−0.6786	0.0687	-	-

Table 5 shows the adjusted R^2 value, standard error, and average relative error in the prediction for the proposed regression curves. For both equations, the R^2 value approaches unity, while the average error is within 4.61%. The F-significances of all the models approach zero.

Table 5. Statistical parameters of the proposed regression models.

Equation	Adjusted R^2	Standard Error	F-Significance	Average Rel. Error (%)
Design model #1	0.9979	27.44 kW	2.2×10^{-180}	1.94
Design model #2	0.9978	29.09 kW	3.5×10^{-179}	1.76
Off-design model	0.9881	0.0251 (-)	0	4.61

For the three regression models, the mean of the residuals is below 1×10^{-13}, hence validating the assumption that the mean of the residuals is close to zero. In addition, the residuals appeared to follow a normal distribution, except for the presence of some tails (see Figure 6). The computed R^2 obtained with a straight trend line are 89.7%, 92.7%, and 96.1%, for the three models, respectively.

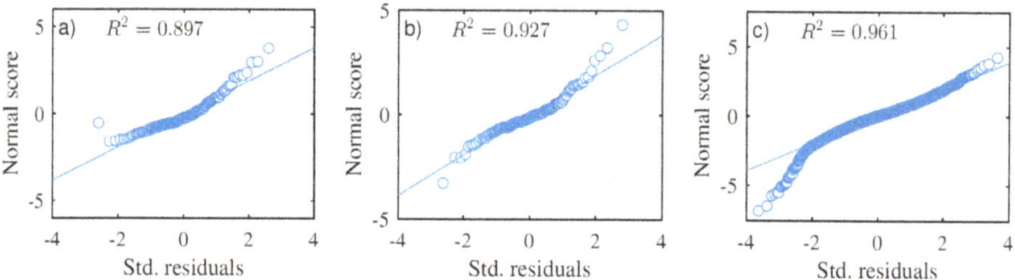

Figure 6. Normal probability plot of the standard residuals: (**a**) design regression model #1, (**b**) design regression model #2, (**c**) off-design regression model (R = 0.897; 0.960).

Figures 7–9 show the scatter plots of the regression's standard residuals as a function of the prediction and the regression parameters (predictors). The plots indicate that the homoscedasticity assumption (the variance of the residuals should not vary as a function of the predicted value, nor the predictor) can be considered acceptable for the off-design regression curve (Figure 9), but is possibly not valid/less convincing in the case of the design point regression models (Figures 7 and 8).

The figures indicate that the spread of the residuals increases, for example, as the predicted ORC net power output increases. A further analysis of the residuals suggests that the absolute error in the prediction increases as a function of the ORC net power output, while the relative deviation remains mostly constant. According to the indications from Gujarati and Porter [37], a violation of the homoscedasticity assumption does not lead to the attainment of biased regression parameters, but rather to an inaccurate estimation of the standard errors of such parameters.

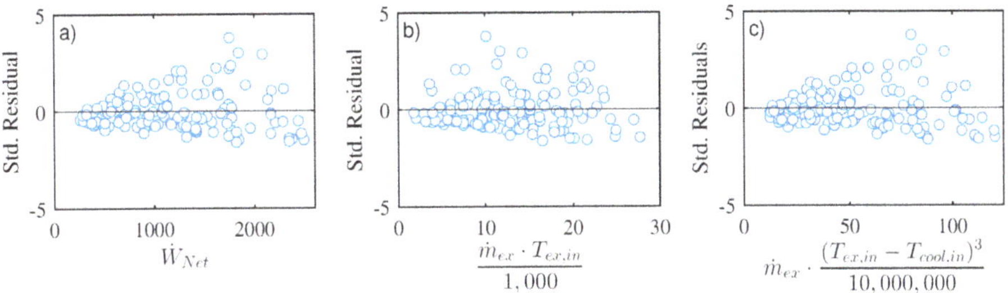

Figure 7. Design regression model #1, standard residuals distribution according to the target data and regression parameters.

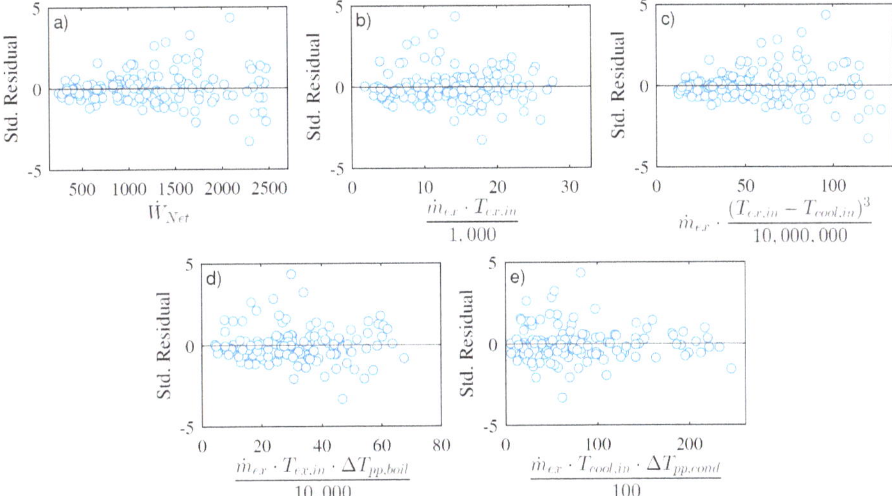

Figure 8. Design regression model #2, standard residuals distribution according to the target data and regression parameters.

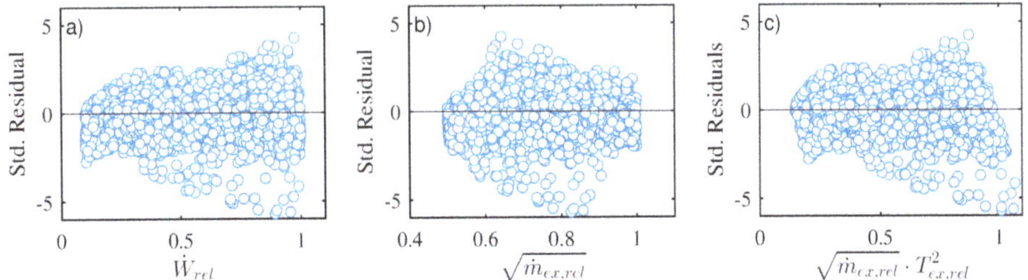

Figure 9. Off-design regression model, standard residuals distribution according to the target data and regression parameters.

Figure 10 depicts the predicted values against the data used for the regression for the two cases and illustrates the fit between data and predictions. The design point models appear to predict the thermodynamic model in a very accurate way, while there is a larger spread of the results in the off-design model, due to the complexity of the phenomena taking place during off-design operation of the unit.

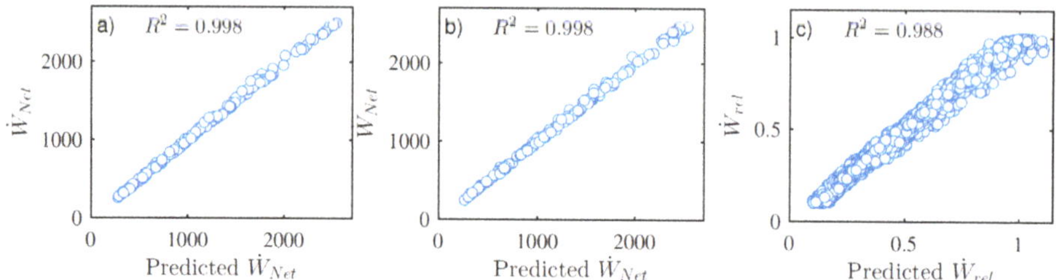

Figure 10. Predicted values against regression data: (**a**) design regression model #1, (**b**) design regression model #1, (**c**) off-design regression model.

3.2. Comparison with the Thermodynamic Evaluations (Model #1—fixed pinch points)

The results of the overall ORC optimization using the thermodynamic simulation model and the regression curves (model #1) are shown in Tables 6 and 7. The results indicate that the simplified approach based on the regression model #1 accurately predicts the optimal design and economic performance of the ORC unit on board the two considered vessels. The maximum deviations are 6.54%, 9.76%, 8.53%, and 6.05% for the ORC design point, ORC design power output, annual production, and LCOE, respectively.

Table 6. Comparison between the solutions attained using the thermodynamic models and the regression curves (model #1): energy production maximization cases.

	Thermodynamic Models	Regression Curves	Difference (%)
Feeder (T_{sw} = 10 °C)			
ORC design load (%)	90	90.0	0.00
ORC design power (kW)	658.7	674.6	2.41
ORC annual production (MWh)	2246	2326	3.56
LCOE ($/kWh)	0.092	0.090	−1.75
Feeder (T_{sw} = 15 °C)			
ORC design load (%)	90	90.0	0.00
ORC design power (kW)	621.8	639.4	2.83
ORC annual production (MWh)	2115	2205	4.21
LCOE ($/kWh)	0.093	0.091	−2.04
Feeder (T_{sw} = 20 °C)			
ORC design load (%)	90	90.0	0.00
ORC design power (kW)	586.6	605.6	3.24
ORC annual production (MWh)	1989	2088	4.97
LCOE ($/kWh)	0.095	0.093	−2.53
Container vessel (T_{sw} = 10 °C)			
ORC design load (%)	100	100.0	0.00
ORC design power (kW)	1500.7	1538.2	2.50
ORC annual production (MWh)	4648	4419	−4.93
LCOE ($/kWh)	0.081	0.086	7.06
Container vessel (T_{sw} = 15 °C)			
ORC design load (%)	100	100.0	0.00
ORC design power (kW)	1409.4	1453.2	3.11
ORC annual production (MWh)	4333	4175	−3.65
LCOE ($/kWh)	0.083	0.088	6.05
Container vessel (T_{sw} = 20 °C)			
ORC design load (%)	100	100.0	0.00
ORC design power (kW)	1323	1372	3.68
ORC annual production (MWh)	4031	3941	−2.22
LCOE ($/kWh)	0.085	0.089	4.95

Table 7. Comparison between the solutions attained using the thermodynamic models and the regression curves (model #1): energy production maximization cases.

	Thermodynamic Models	Regression Curves	Difference (%)
Feeder (T_{sw} = 10 °C)			
ORC design load (%)	75	78.4	4.47
ORC design power (kW)	458.2	498.0	8.69
ORC annual production (MWh)	1843	1979	7.40
LCOE ($/kWh)	0.086	0.085	−1.05
Feeder (T_{sw} = 15 °C)			
ORC design load (%)	75	78.4	4.48
ORC design power (kW)	430.7	470.3	9.20
ORC annual production (MWh)	1732	1869	7.93
LCOE ($/kWh)	0.087	0.086	−1.15
Feeder (T_{sw} = 20 °C)			
ORC design load (%)	75	78.4	4.47
ORC design power (kW)	404.2	443.7	9.76
ORC annual production (MWh)	1625	1763	8.53
LCOE ($/kWh)	0.089	0.087	−1.80
Container vessel (T_{sw} = 10 °C)			
ORC design load (%)	37.45	35.0	−6.54
ORC design power (kW)	633.9	631.5	−0.38
ORC annual production (MWh)	3831	3823	−0.20
LCOE ($/kWh)	0.0522	0.0522	0.00
Container vessel (T_{sw} = 15 °C)			
ORC design load (%)	35.73	35.0	−2.05
ORC design power (kW)	587.7	596.5	1.49
ORC annual production (MWh)	3560	3611	1.44
LCOE ($/kWh)	0.053	0.053	−0.38
Container vessel (T_{sw} = 20 °C)			
ORC design load (%)	35	35.0	0.00
ORC design power (kW)	547.8	562.9	2.76
ORC annual production (MWh)	3318	3408	2.70
LCOE ($/kWh)	0.0542	0.054	−0.74

From the perspective of the computational time, the thermodynamic optimization requires around 1 hour of simulation time, while the simplified approach requires less than 1 second. The ORC evaporation temperatures were found to be in the range from 137.6 to 160.8 °C, while the ORC condensation temperature were in the range from 22.4 to 32.4 °C.

Figure 11 shows the computed ORC power production as a function of the main engine load for the considered cases.

Even in this case, the regression models are proven to be capable of accurately reproducing the part load operation of the optimized ORC units along the various engine loads. The greatest deviations appear in Figure 10c (feeder optimized for minimum LCOE), where the ORC power production is slightly overestimated when the engine is operated at loads above 60%. This is due to an overestimation of the ORC design point, which is found to be at an engine load of 75% in the thermodynamic simulations, and at an engine load of 78.4% when using the regression models.

Regarding the optimal design of ORC units, finally, it emerges that units that are optimized to minimize the LCOE are designed for lower engine loads, compared to units that are optimized to maximize the overall energy production. In particular, the design load that minimizes the system LCOE is of 75% and 35% for the feeder and the container vessel, respectively. Designing the unit for a lower engine load ensures that the ORC is operated in the design point for a higher amount of time, and hence a lower cost of the produced energy, compared to when the unit is mostly operated in part-load conditions, with lower conversion efficiencies.

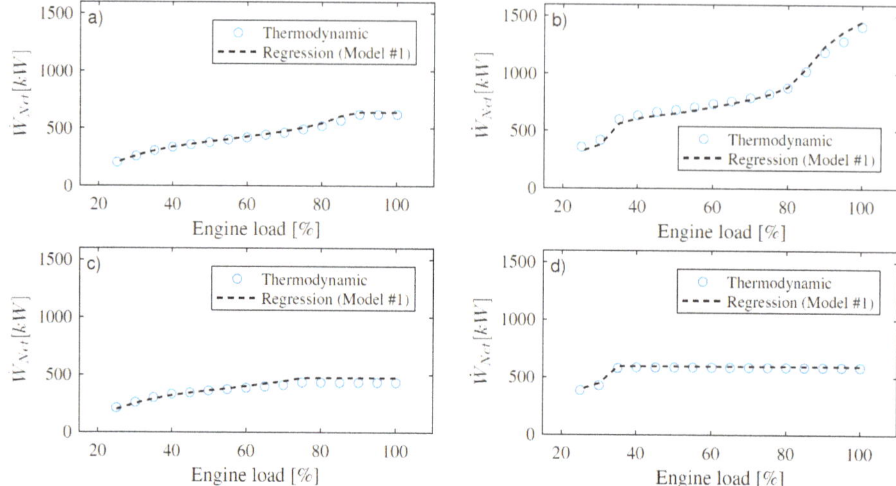

Figure 11. ORC power output as a function of the engine load estimated by the thermodynamic model and by the regression model #1: (**a**) feeder—maximization of energy production, (**b**) container vessel—maximization of energy production, (**c**) feeder—minimization of LCOE, (**d**) container vessel—minimization of LCOE.

3.3. Comparison with the Thermodynamic Evaluations (Model #2—variable pinch points)

Tables 8 and 9 show the results of the overall ORC optimizations using the thermodynamic simulation tool and the regression model #2. In all cases, a sea water temperature of 15 °C was assumed. Table 8 displays the solutions attained when maximizing the ORC annual energy production, while Table 9 depicts the results attained when minimizing the LCOE of the system. Different cases were considered, where the minimum pinch points temperatures both in the boiler and the condenser are varied. This allows to check the suitability of the regression model #2 to capture the impact of the minimum pinch point temperatures on the attainable performance of the unit. The results suggest that the maximum deviations for the engine load for which the ORC should be designed, ORC design power output, annual production, and LCOE, are 4.47%, 11.14%, 10.01%, and 6.71%. As it could be expected, the inclusion of more model parameters, namely the pinch points, decreases the accuracy of the estimations. The average deviations in the estimated annual productions and LCOE are 4.48% and 2.5%, indicating that even regression model #2 is suitable to carry out preliminary estimations for the optimal design and performance of ORC units to be installed on board ships using low-sulfur fuels.

Table 8. Comparison between the solutions attained using the thermodynamic models and the regression curves (model #2): energy production maximization cases.

	Thermodynamic Models	Regression Curves	Difference (%)
Feeder ($\Delta T_{pp,boil} = 25\,°C$; $\Delta T_{pp,cond} = 10\,°C$)			
ORC design load (%)	90.0	90.0	0.0
ORC design power (kW)	575.2	609.5	6.0
ORC annual production (MWh)	1961	2102	7.2
LCOE ($/kWh)	0.095	0.093	−2.6
Feeder ($\Delta T_{pp,boil} = 20\,°C$; $\Delta T_{pp,cond} = 8\,°C$)			
ORC design load (%)	90.0	90.0	0.00
ORC design power (kW)	621.8	640.4	2.98
ORC annual production (MWh)	2115	2208	4.37
LCOE ($/kWh)	0.093	0.091	−2.15
Feeder ($\Delta T_{pp,boil} = 15\,°C$; $\Delta T_{pp,cond} = 5\,°C$)			
ORC design load (%)	90.0	90.0	0.00
ORC design power (kW)	678.9	673.1	−0.85
ORC annual production (MWh)	2300	2321	0.90
LCOE ($/kWh)	0.091	0.090	−1.53
Container vessel ($\Delta T_{pp,boil} = 25\,°C$; $\Delta T_{pp,cond} = 10\,°C$)			
ORC design load (%)	100.0	100.0	0.00
ORC design power (kW)	1300.0	1380.6	6.20
ORC annual production (MWh)	4050	3996	−1.32
LCOE ($/kWh)	0.083	0.089	6.71
Container vessel ($\Delta T_{pp,boil} = 20\,°C$; $\Delta T_{pp,cond} = 8\,°C$)			
ORC design load (%)	100.0	100.0	0.00
ORC design power (kW)	1409.4	1460.6	3.63
ORC annual production (MWh)	4333	4196	−3.15
LCOE ($/kWh)	0.083	0.088	5.93
Container vessel ($\Delta T_{pp,boil} = 15\,°C$; $\Delta T_{pp,cond} = 5\,°C$)			
ORC design load (%)	100.0	100.0	0.00
ORC design power (kW)	1542.0	1545.9	0.25
ORC annual production (MWh)	4656	4441	−4.62
LCOE ($/kWh)	0.082	0.086	4.99

It emerges that the optimal design load of the engine for which the ORC should be designed is not affected by the selected pinch point values. Additionally, as expected, a decrease in the pinch points is connected with an increase of the annual energy production from the ORC. The ORC production curves as a function of the engine load are not reported for this case, because a comparison between the results attained using model #1 and model #2 for the case where the pinch points are set to 20 °C (for the boiler) and 8 °C (for the condenser), indicated that the power production curves given by the two regression models are overlapping. The ORC evaporation temperatures were found to be in the range from 137.5 to 162.4 °C, while the ORC condensation temperatures were in the range from 24.4 to 29.5 °C.

Table 9. Comparison between the solutions attained using the thermodynamic models and the regression curves (model #2): LCOE minimization cases.

	Thermodynamic Models	Regression Curves	Difference (%)
Feeder ($\Delta T_{pp,boil} = 25\ °C; \Delta T_{pp,cond} = 10\ °C$)			
ORC design load (%)	75.0	78.4	4.28
ORC design power (kW)	397.1	446.9	11.14
ORC annual production (MWh)	1598	1776	10.01
LCOE ($/kWh)	0.089	0.087	−1.95
Feeder ($\Delta T_{pp,boil} = 20\ °C; \Delta T_{pp,cond} = 8\ °C$)			
ORC design load (%)	75.0	78.4	4.47
ORC design power (kW)	430.7	473.8	10.02
ORC annual production (MWh)	1732	1883	8.74
LCOE ($/kWh)	0.087	0.086	−1.38
Feeder ($\Delta T_{pp,boil} = 15\ °C; \Delta T_{pp,cond} = 5\ °C$)			
ORC design load (%)	75.0	78.4	4.47
ORC design power (kW)	474.3	502.6	5.95
ORC annual production (MWh)	1902	1997	5.02
LCOE ($/kWh)	0.085	0.085	−0.71
Container vessel ($\Delta T_{pp,boil} = 25\ °C; \Delta T_{pp,cond} = 10\ °C$)			
ORC design load (%)	35.5	35.0	−1.39
ORC design power (kW)	539.8	566.8	4.77
ORC annual production (MWh)	3273	3431	4.63
LCOE ($/kWh)	0.054	0.054	−1.12
Container vessel ($\Delta T_{pp,boil} = 20\ °C; \Delta T_{pp,cond} = 8\ °C$)			
ORC design load (%)	35.7	35.0	−2.05
ORC design power (kW)	587.7	600.4	2.15
ORC annual production (MWh)	3560	3635	2.11
LCOE ($/kWh)	0.053	0.053	−0.56
Container vessel ($\Delta T_{pp,boil} = 15\ °C; \Delta T_{pp,cond} = 5\ °C$)			
ORC design load (%)	36.5	35.0	−4.26
ORC design power (kW)	649.2	636.3	−2.03
ORC annual production (MWh)	3922	3852	−1.81
LCOE ($/kWh)	0.052	0.052	0.38

4. Discussion

The procedure followed to derive the proposed regression curves is typical of the data-driven modelling approach, which is now gaining interest due to the increasing data availability and the development of more and more sophisticated machine-learning algorithms. The proposed algorithms are attained by implementing a supervised machine learning approach, where both the model outputs (i.e., the ORC power output) and the model input (i.e., the characteristics of the main engine exhaust gases) are known.

A more rigorous implementation of the data-driven modelling approach to derive the regression curves would have required to split the available dataset into training and test sets. This is because regression algorithms are generally developed based on a portion of the overall data (train set), and then their accuracy is tested on the remaining portion of the data (test set). This allows to prove the reliability of the proposed algorithm to predict the model output also for data outside the space covered by the data used for its development.

This was considered not to be required for our case, because the accuracy of the proposed approach, which combines design and off-design regression models, was verified through several case studies, whose inputs parameters were not used when generating the regression curves.

In any case, the authors do not recommend the use of the proposed regression models when using such curves outside their validation space (i.e., increased temperatures of the exhaust gases, or reduced flow rates).

In addition, as a way to reduce the model complexity, several assumptions were made which could influence the attained results. In particular, the evaluations were limited to one fluid, the backpressure effect on the engine was neglected, and the pressure drops in the heat exchangers were not accounted for. The decision not to consider the impact on the backpressure effect was mainly connected to two reasons: (1) the findings from Michos et al. [14] indicated that such parameter has a weak impact on the overall estimated savings, and (2) the acceptable backpressure level to the engine varies both with the selected engine and the considered engine load, making it a complex feature to include in simplified regression models.

With respect to the approach used to estimate the cost of the unit, it should be mentioned that large uncertainties are expected, because the cost of the unit is not uniquely identified by its overall size, but is influenced also by other parameters, such as the operating pressures, the selected materials, and the configuration of the installation in the engine system. The attained economic results are therefore to be considered as preliminary and are meant to give a first estimate.

5. Conclusions

This paper presented an accurate and time-efficient method to estimate the techno-economic prospects for organic Rankine cycle-based waste heat recovery units tailored for ships using low-sulfur fuels. The proposed method is derived by integrating regression curves capable of describing the design and off-design performance of organic Rankine cycle units. The user-specified input parameters are the temperature and mass flow rate of the ship's exhaust gases, as well as the ship sailing profile. Both the statistical significance and the expected accuracy of the proposed regression curves were analyzed. In comparison with the evaluations carried out using standard thermodynamic-based models, maximum deviations of 8.53% and 6.05% in the estimated organic Rankine cycle annual energy production and levelized cost of electricity were attained when considering the regression model #1 (fixed ORC pinch points). The deviations increased to 10% for the annual energy production, and 6.7% for the levelized cost of electricity for the regression model #2, when the unit's pinch points were included as parameters in the regression curves. The proposed method assumes that the organic Rankine cycle that maximizes the net power production in design conditions is the one resulting in the greatest energy production on an annual basis. The comparison between the regression models and the thermodynamic evaluations, which did not include this premise, demonstrated that this assumption does not significantly affect the accuracy of the estimations.

The proposed method is simple, leads to accurate estimations, and is characterized by a short computational time (less than one second). This makes it suitable to be used for preliminary evaluations of the potential for installing organic Rankine cycle power systems on board a wide range of vessel types. In addition, the proposed approach can be used even by non-experts in the organic Rankine cycle technology, and hence it can be of interest for a wide audience of readers.

Author Contributions: Conceptualization, E.B., M.M., U.L, and F.H.; methodology E.B. and M.M.; software, E.B.; formal analysis, E.B.; writing—original draft preparation, E.B.; writing—review and editing, E.B., M.M., U.L, and F.H.; visualization, E.B.; supervision; M.M., U.L, and F.H; Funding acquisition F.H.; Project administration F.H. All authors have read and agreed to the published version of the manuscript.

Funding: Enrico Baldasso carried out the work within the frames of the project "Waste recovery on liquefied natural gas-fueled ships", funded by Orients Fond and Den Danske Maritime Fond.

Conflicts of Interest: The authors declare no conflict of interest.

Nomenclature

Acronyms

CEPCI	Chemical Engineering Plant Cost Index
CO_2	carbon dioxide
E	Electricity generation
GHG	Greenhouse gases
HFO	heavy fuel oil
IMO	International Maritime Organization
LNG	liquefied natural gas
NO_x	nitrogen oxides
SO_x	sulfur oxides
WHR	waste heat recovery

Symbols

E	electricity generation, kWh
I	investment cost, US $
LCOE	levelized cost of electricity, $/kWh
O&M	operation and maintenance
P	pressure, bar
r	discount rate
$\dot{W}$	power, kW

Subscripts and superscripts

crit	critical
des	design
ex	exhaust
exp	expander
gear	gearbox
gen	generator
in	inlet
off	off-design
p	pump
rel	relative
sw	seawater
y	year

References

1. *Nitrogen Oxides (NOx) Regulation 13*; Tech. Rep.; International Maritime Organization: London, UK, 2015.
2. *Sulfur Oxides (SOx) Regulation 14*; Tech. Rep.; International Maritime Organization: London, UK, 2015.
3. The International Maritime Organization. *Adoption of the Initial IMO Strategy on Reduction of GHG Emissions from Ships and Existing IMO Activity Related to Reducing GHG Emissions in the Shipping Sector*; Tech Rep; International Maritime Organization: London, UK, 2018.
4. Brynolf, S.; Fridell, E.; Andersson, K. Environmental assessment of marine fuels: Liquefied natural gas, liquefied biogas, methanol and bio-methanol. *J. Clean. Prod.* **2014**, *74*, 86–95. [CrossRef]
5. Smith, A.B. Gas fuelled ships: Fundamentals, benefits classification & operational issues. In Proceedings of the 1st Gas Fuelled Conference, Hamburg, Germany, 21 October 2010.
6. Bouman, E.A.; Lindstad, E.; Rialland, A.I.; Strømman, A.H. State-of-the-art technologies, measures, and potential for reducing GHG emissions from shipping—A review. *Transp. Res. Part D* **2017**, *52*, 408–421. [CrossRef]
7. Baldi, F.; Ahlgren, F.; Nguyen, T.; Thern, M.; Andersson, K. Energy and exergy analysis of a cruise ship. *Energies* **2018**, *11*, 1–41. [CrossRef]

8. Baldi, F.; Nguyen, T.; Ahlgren, F. The application of process integration to the optimisation of cruise ship energy systems: A case study In a context of demand for increased energy efficiency in shipping, this article proposes an example. In Proceedings of the ECOS 2016: 29th International Conference on Efficiency, Cost, Optimization, Simulationand Envirionmental Impact of Energy Systems, Portoroz, Slovenia, 19–23 June 2016.

9. Mondejar, M.E.; Andreasen, J.G.; Pierobon, L.; Larsen, U.; Thern, M.; Haglind, F. A review on the use of organic Rankine cycle power systems for marine applications. *Renew. Sustain. Energy Rev.* **2018**, *91*, 126–151. [CrossRef]

10. Larsen, U.; Sigthorsson, O.; Haglind, F. A comparison of advanced heat recovery power cycles in a combined cycle for large ships. *Energy* **2014**, *74*, 260–268. [CrossRef]

11. Andreasen, J.G.; Meroni, A.; Haglind, F. A comparison of organic and steam Rankine cycle power systems for waste heat recovery on large ships. *Energies* **2017**, *10*, 1–23. [CrossRef]

12. Soffiato, M.; Frangopoulos, C.A.; Manente, G.; Rech, S.; Lazzaretto, A. Design optimization of ORC systems for waste heat recovery on board a LNG carrier. *Energy Convers. Manag.* **2015**, *92*, 523–534. [CrossRef]

13. Baldi, F.; Larsen, U.; Gabrielii, C. Comparison of different procedures for the optimisation of a combined Diesel engine and organic Rankine cycle system based on ship operational profile. *Ocean Eng.* **2015**, *110*, 85–93. [CrossRef]

14. Michos, C.N.; Lion, S.; Vlaskos, I.; Taccani, R. Analysis of the backpressure effect of an Organic Rankine Cycle (ORC) evaporator on the exhaust line of a turbocharged heavy duty diesel power generator for marine applications. *Energy Convers. Manag.* **2017**, *132*, 347–360. [CrossRef]

15. Baldasso, E.; Andreasen, J.G.; Mondejar, M.E.; Larsen, U.; Haglind, F. Technical and economic feasibility of organic Rankine cycle-based waste heat recovery systems on feeder ships: Impact of nitrogen oxides emission abatement technologies. *Energy Convers. Manag.* **2019**, *183*, 577–589. [CrossRef]

16. Rech, S.; Zandarin, S.; Lazzaretto, A.; Frangopoulos, C.A. Design and off-design models of single and two-stage ORC systems on board a LNG carrier for the search of the optimal performance and control strategy. *Appl. Energy* **2017**, *204*, 221–241. [CrossRef]

17. Liu, B.T.; Chien, K.H.; Wang, C.C. Effect of working fluids on organic Rankine cycle for waste heat recovery. *Energy* **2004**, *29*, 1207–1217. [CrossRef]

18. Kuo, C.R.; Hsu, S.W.; Chang, K.H.; Wang, C.C. Analysis of a 50 kW organic Rankine cycle system. *Energy* **2011**, *36*, 5877–5885. [CrossRef]

19. Wang, D.; Ling, X.; Peng, H.; Liu, L.; Tao, L.L. Efficiency and optimal performance evaluation of organic Rankine cycle for low grade waste heat power generation. *Energy* **2013**, *50*, 343–352. [CrossRef]

20. Larsen, U.; Pierobon, L.; Wronski, J.; Haglind, F. Multiple regression models for the prediction of the maximum obtainable thermal ef fi ciency of organic Rankine cycles. *Energy* **2013**, 1–8. [CrossRef]

21. Lecompte, S.; Huisseune, H.; van den Broek, M.; De Paepe, M. Methodical thermodynamic analysis and regression models of organic Rankine cycle architectures for waste heat recovery. *Energy* **2015**, *87*, 60–76. [CrossRef]

22. Palagi, L.; Sciubba, E.; Tocci, L. A neural network approach to the combined multi-objective optimization of the thermodynamic cycle and the radial inflow turbine for Organic Rankine cycle applications. *Appl. Energy* **2019**, *237*, 210–226. [CrossRef]

23. Dickes, R.; Dumont, O.; Quoilin, S.; Lemort, V. Performance correlations for characterizing the optimal off-design operation of an ORC power system. *Energy Procedia* **2017**, *129*, 907–914. [CrossRef]

24. Baldasso, E.; Elg, M.; Haglind, F.; Baldi, F. Comparative Analysis of Linear and Non-Linear Programming Techniques for the Optimization of Ship Machinery Systems. *J. Mar. Sci. Eng.* **2019**, *7*. [CrossRef]

25. Andreasen, J.G.; Larsen, U.; Knudsen, T.; Pierobon, L.; Haglind, F. Selection and optimization of pure and mixed working fluids for low grade heat utilization using organic rankine cycles. *Energy* **2014**, *73*, 204–213. [CrossRef]

26. Rayegan, R.; Tao, Y.X. A procedure to select working fluids for Solar Organic Rankine Cycles (ORCs). *Renew. Energy* **2011**, *36*, 659–670. [CrossRef]

27. Drescher, U.; Brüggemann, D. Fluid selection for the Organic Rankine Cycle (ORC) in biomass power and heat plants. *Appl. Therm. Eng.* **2007**, *27*, 223–228. [CrossRef]

28. MAN Energy Solutions. *Waste Heat Recovery Systems (WHRS)—Marine Engines & Systems*; Tech. Rep.; MAN Energy Solutions: Copenhagen, Denmark, 2014.

29. Bell, I.H.; Wronski, J.; Quoilin, S.; Lemort, V. Pure and pseudo-pure fluid thermophysical property evaluation and the open-source thermophysical property library CoolProp. *Ind. Eng. Chem. Res.* **2014**, *53*, 2498–2508. [CrossRef] [PubMed]

30. MathWorks Official Website. Available online: https://mathworks.com/products/matlab.html (accessed on 31 January 2018).

31. Haglind, F.; Mondejar, M.E.; Andreasen, J.G.; Pierobon, L.; Meroni, A. *Organic Rankine Cycle Unit for Waste Heat Recovery on Ships (PilotORC)—Final Report*; Tech. Rep.; Department of Mechanical Engineering, Technical University of Denmark (DTU): Copenhagen, Denmark, 2017.

32. Usman, M.; Imran, M.; Yang, Y.; Hyun, D.; Park, B. Thermo-economic comparison of air-cooled and cooling tower based Organic Rankine Cycle (ORC) with R245fa and R1233zde as candidate working fl uids for different geographical climate conditions. *Energy* **2017**, *123*, 353–366. [CrossRef]

33. Turton, R.; Bailie, R.C.; Whiting, W.B.; Shaeiwitz, J.A.; Bhattacharyya, D. *Analysis, Synthesis and Design of Chemical Processes*, 3rd ed.; Pearson Education: London, UK, 2013; ISBN 9788578110796.

34. Lemmens, S. Cost engineering techniques and their applicability for cost estimation of organic rankine cycle systems. *Energies* **2016**, *9*, 485. [CrossRef]

35. Sobol, I.M.; Turchaninov, V.; Levitan, Y.L.; Shukhamn, B.V. Quasirandom sequence generators. In Proceedings of the IPM ZAK, no. 30; Keldysh Institute of Applied Mathematics, Russian Academy of Sciences: Moscow, Russia, 1992.

36. MAN Energy Solutions CEAS Calculation Tool. Available online: https://marine.man-es.com/two-stroke/ceas (accessed on 12 December 2018).

37. Guijarati, D.N.; Porder, D.C. *Basic Econometrics*, 6th ed.; McGraw-Hill: Boston, MA, USA,, 2009.

Article

Life Cycle Assessment of a Commercially Available Organic Rankine Cycle Unit Coupled with a Biomass Boiler

Anna Stoppato and Alberto Benato *

Department of Industrial Engineering, University of Padova, 35131 Padova, Italy; anna.stoppato@unipd.it
* Correspondence: alberto.benato@unipd.it; Tel.: +39-049-827-6752

Received: 7 March 2020; Accepted: 4 April 2020; Published: 10 April 2020

Abstract: Organic Rankine Cycle (ORC) turbogenerators are a well-established technology to recover from medium to ultra-low grade heat and generate electricity, or heat and work as cogenerative units. High firmness, good reliability and acceptable efficiency guarantee to ORCs a large range of applications: from waste heat recovery of industrial processes to the enhancement of heat generated by renewable resources like biomass, solar or geothermal. ORC unit coupled with biomass boiler is one of the most adopted arrangements. However, despite biomass renewability, it is mandatory to evaluate the environmental impact of systems composed by boilers and ORCs taking into account the entire life cycle. To this purpose, the authors perform a life cycle assessment of a commercially available 150 kW cogenerative ORC unit coupled with a biomass boiler to assess the global environmental performance. The system is modelled in SimaPro using different approaches. Results show that the most impacting processes in terms of CO_2 equivalent emissions are the ones related to biomass production and organic fluid leakages with 71% and 19% of the total. Therefore, being fluid release in the environment high impacting, a comparison among three fluids is also performed. Analysis shows that adopting a hydrofluoroolefin fluid with a low global warming potential instead of the hydrocarbon fluid as already used in the cycle guarantees a significant improvement of the environmental performance.

Keywords: life cycle assessment; ORC; biomass; CHP; carbon footprint of energy production

---

1. Introduction

In 2018, despite the around \$280 Billion of global investment in renewable energy sources (RES) [1], global energy-related carbon dioxide (CO_2) emissions rose by 1.6% [2]. An issue source of great concerns for people involved in the development of new and more eco-friendly energy scenarios.

In fact, at the time of writing, fossil fuels accounted for more than 85% in the coverage of the world primary energy consumption: 11,485 out of 13,511.2 Mtoe [1].

However, based on IRENA estimations [3], to meet the 2 °C climate goal, RES share in final energy consumption needs to rise from the 19% registered in 2017, to 65% in 2050. This means a constant and massive trend of growth for RES in the upcoming years.

Nevertheless, there would seem to be a need for general clarification. First of all, the analysis of 2018 available data [1], reveals an extremely interesting point: 70% of new power additions were of renewable origin.

These RES installations were driven by the sky-rocket growth in terms of installed capacity of wind and solar: +21% and +47%, respectively. A fact that pushed RES coverage of the world primary energy consumption over 10%.

However, the growth in RES installations is not random; it is mainly driven by massive support actions. Indeed, starting from the year 2009, renewable energy sources are the subject of several

European Union (EU) economic supports [4–6] because RES are considered the most promising way to reduce both primary energy consumption and CO_2 emissions.

Thanks to these actions, EU primary energy consumption decreased from 1823.9 Mtoe to 1689.2 Mtoe in the period 2007–2017 while CO_2 emissions drooped down of 9.4% in the period 2005–2016. These optimistic numbers confirm the effectiveness of these measures especially in view of a transition to RES based power generation systems characterized by massive electrification of transportation, buildings and industry.

Notwithstanding wind and solar eco-friendliness, the widespread availability and their undoubted considerable potential, it is important to remark that these sources are difficult to predict and exhibit large variation in space and time, characteristics which give rise to large unforeseeable power fluctuations. However, being electricity demand instantaneously balanced 24/7, huge and unpredictable power changes can be the source of imbalances between supply and demand that can provoke grid damage, user devices fault or blackouts. These unpleasant effects can be managed and/or limited using fossil fuels power plants working in cycling operation mode and energy storage units (see e.g., [7–17]).

Note that wind and solar variability and, consequently, unpredictability can be greatly affected by climate change because they are directly linked to climate variables like solar irradiation, wind, temperature and precipitation. So, being that the reliability and performance of RES plants are largely affected by climate conditions, it is also mandatory to consider climate change future impacts on meteorological conditions during the design of renewable-based energy systems. This approach can avoid mistakes during the design of energy infrastructures: systems characterized by high investment costs, long lifespan and a not negligible environmental impact.

Leaving aside concerns over climate change future impacts on RES based energy systems as well as solar and wind great potential and remarkable variability and unforeseeability, it is important to point out that there is a renewable source with great potential and a programmable production, able to generate heat, electricity or a combination of them, but first and foremost that largely contributes to gross final energy consumption and total primary energy supply. This RES is biomass.

Analyzing the latest data referred to the year 2016 [18], RES contribution to total primary energy supply is 14%, of which about 70% from biomass (9.8% on the total primary energy), while it is 17.9% related to the gross final energy consumption of which about 73% from biomass (13% on the total gross final energy).

Therefore, it is a matter of fact that biomass is a key source due to its widespread availability, production programmability and capability of generating electricity, heat or a combination of them.

Among RES, the electricity production coming from biomass ranks third after hydro and wind, while, for heat generation, biomass is the undisputed leader among renewables with about 96% of production.

A deeper look at biomass data [18] clearly shows that solid biomass with over 65% is the main contributor to electricity production, as well as of heat generation. This key role in both generation sectors is already visible for both developed and developing regions and can become crucial in the future to push the transition towards a decarbonized energy sector and a wider availability of energy.

Concerning electricity generation using biomass, several technologies are available according to the biomass type. As an example, liquid biofuels or biogas can be burnt in an Internal Combustion Engine (ICE) or in a Gas Turbine (GT) to generate electricity only or electricity and heat in Combined Heat and Power (CHP) units (see, e.g., [19–25]). In these cases, plant arrangements are really simple because the ICE and/or the GT, being their shaft mechanically coupled with the electric generator, directly produce electricity while heat is extracted from ICE coolant and lube oil or from the GT exhaust gases.

On the contrary, if solid biomass or municipal/industrial solid wastes are used as biofuel, the plant arrangement is more complex.

In fact, when electricity is generated from solid biomass like woodchip, pellet, ect., plant architecture comprises a boiler in which the combustion process takes place and a Waste Heat Recovery Unit (WHRU) devoted to recovering the exhaust gases heat content and convert it into electricity or, if requires, into electricity and heat.

In literature, several WHRU types are available but the most adopted technologies are Steam Rankine Cycles—SRCs for large size units and Organic Rankine Cycles—ORCs for medium and low power unit (usually with an electric power lower than 2 MW).

Since the vast majority of applications is characterized by medium to low power, the conventional plant arrangement that adopts solid biomass as fuel is composed by a boiler and an ORC turbogenerator.

The organic Rankine cycle technology has been investigated since the years 1880, but its popularity increased only in the years 2000 with the growing interest in exploiting from medium to ultra-low grade heat sources. In fact, when the heat source temperature ranges from 80 °C to 500 °C, conventional steam Rankine cycle units fail for technical and economical reasons (see, e.g., [26–28]. Therefore, to recover the heat of these sources and convert it into useful electrical energy or electricity and heat, the ORC technology is the most efficient option.

An ORC operates in the same way as a steam Rankine cycle but an organic compound (hydrocarbons, hydrofluorocarbons, hydro-chlorofluorocarbons, chlorofluorocarbons, per-fluorocarbons, siloxanes, alcohols, aldehydes, ethers hydrofluoroethers, amines, zeotropic and azeotropic mixtures) is used as working fluid instead of water.

Due to the availability of a large number of possible working fluids and several plant schemes (basic, regerative, recuperative and regenerative, etc. and with or without intermediate oil or water loop—for more details see, e.g., [27]), the ORC unit design is a complicated task.

For this reason, thousands of researches have been conducted around the world with the aim of selecting the best working fluid and plant configuration for a wide range of sources as well as optimizing and testing plant or its components (see, e.g., [29–37]). There are also works devoted to environmental and exergetic assessment of power generation units.

However, despite the large variety of published works, there is a lack of studies aiming to quantify the environmental impact of an in-operation organic Rankine cycle power plant coupled with a biomass boiler. To perform this kind of investigation, the Life Cycle Assessment (LCA) approach needs to be adopted.

In a nutshell, Life Cycle Assessment is today one of the most accredited assessment methods at the international level for the quantification of damage and its outcomes because these can be immediately correlated to impacts on human health, quality of ecosystems and consumption of natural resources [38].

The life cycle analysis allows to go beyond the geographical boundaries of the plant and to assess the impacts throughout the "value chain", such as those associated with emissions due to the entire biomass supply chain as well as construction, operation and decommissioning phases of plants.

Note that, Life Cycle Assessment has already been applied to biomass power generation unit or ORC. As an example, Stougie et al. [39] performed an environmental and exergetic sustainability assessment of different plants using biomass. Bioethanol from verge grass and biogas from manure are considered as fuel. Results are focused on the effects of different allocations strategies and underline that the environmental sustainability depends on the way of dealing with by-products.

With respect to the life cycle assessment of ORCs, researchers mainly focus on environmental benefits coming from the installation of ORC units in energy-intensive processes or in RES installations. As an example, Walsh and Thornley [40] evaluated the environmental impact and economic feasibility of introducing an organic Rankine cycle to recover low-grade heat during the production of metallurgical coke. Results show that an ORC can compensate from 1 to 3% of the CO_2 directly emitted during the coke production. This means a yearly reduction of 10,000 t CO_2. In addition, the electricity generated by the ORC replaces all the industry imported electricity. This fact contributes to compute a relatively attractive pay-back period: less than six years. Based on these findings, the authors concluded that ORC can be considered a cost-effective method of achieving Greenhouse Gas (GHG) savings in process industries.

Liu et al. [41] investigated the environmental impact of an organic Rankine cycle power plant for waste heat recovery. The considered heat source temperature and mass flow are 423.15 K and 1 kg/s, respectively, while seven possible working fluids are tested. Results exhibit that the ORC operating with R113 allows the lowest environment impact load followed by the one working with Pentane.

Heberle et al. [42] performed an LCA of organic Rankine cycles for geothermal power generation considering low-GWP working fluids. In particular, the analysis focuses on organic fluids' environmental impact and, specifically, on the impact linked to fluid losses. Results underline that substituting R245fa and R134a with R1233zd, R1234yf or natural hydrocarbons guarantees to significantly reduce the environmental impact. In addition, adopting a two-stage ORC configuration operating with R1233zd instead of a subcritical one-stage system working with R245fa leads to 2% higher exergetic efficiency and a reduction of the global warming impact from 78 g CO_2/kWh to 13 g CO_2/kWh.

Lin et al. [43] used the LCA to evaluate the benefit on environmental impacts introduced by the use of an ORC and wood pellet fuel in the electric arc furnace steel industry. Results reveal that the environmental impacts on ecosystems, human health and resource depletion can be mitigated by adopting the ORC and replacing the heavy oil with wood pellets.

Uusitalo et al. [44] investigated the possibility of reducing GHG emission of biogas engine by the generation of additional electricity from the recovery of waste heat. ORC is the selected WHRU and two scenarios are compared: (i) the ICE generates only electricity and the produced heat is considered a loss, (ii) the exhaust heat from the biogas engine is used as a heat source to generate additional electricity via the ORC technology. LCA reveals that the ORC technology guarantees to reduce the annual GHG emission of at least 280 t $CO_{2,eq}$. So, ORCs can reduce GHG emissions and increase renewable electricity production from biogas engines.

Sedpho et al. [45] performed a conventional and exergetic life cycle assessment on an organic Rankine cycle mounted on a to municipal waste management facility. The selected case study is Mae Hong Son province in Thailand. LCA results highlight that the environmental impact and the energy consumption can be reduced by applying the waste treatment technology of the Refuse-Derived Fuel and a 20 kW organic Rankine cycle unit.

Obviously, several other works dealing with LCA analysis of pellet boilers (e.g., [46]), renewable and fossil fuels (e.g., [47,48]), photovoltaic panels (e.g., [49]), conventional and near-zero energy buildings (e.g., [50]), concert (e.g., [51]), etc. are available in literature. However, as said, to the authors' knowledge, no one has previously presented an investigation devoted to analyzing with the LCA approach a commercially available biomass boiler coupled with a 150 kW cogenerative ORC unit with the aim of understating the system environmental impacts. Being the boiler fed by woodchip, the entire production pathway, starting from wood cultivation, is considered as well as the ORC unit fluid losses along its lifespan. In this regard, in the analysis, three working fluids are tested with the aim of assessing the less impacting one but still maintaining the ORC performance.

The rest of the paper is organized as follows. Section 2 presents the case study and the model setting while in Section 3 materials and methods are described. Section 4 reports the LCA outcomes while in Section 5 conclusion remarks are given.

2. Case Study and Model Settings

The analysis is performed considering a real in-operation system located in the Northeast of Italy. The plant is powered by woodchip of which about 40% is made of birch, 24% of spruce, 20% of pine, 14% of beech and 2% of oak wood. Woodchip lower heating value on dry mass is 18.9 MJ/kg. The timber comes from dedicated woods grown within a radius of 75 km from the plant while its transportation is done by means of freight lorry.

The ORC module, whose main features are summarized in Table 1, is a commercially available Combined Heat and Power unit adopting as working medium a mixture of hydrocarbons (for simplicity the fluid is named "HCF") with a Global Warming Potential (GWP) equal to 950 and an Ozone Depletion Potential (ODP) equal to 0.

The ORC works with a regenerative sub-critical cycle where the useful heat is delivered as hot water at about 80 °C.

A schematic view of the system and its main devices is given in Figure 1 and Table 2.

Table 1. Main features of the Organic Rankine Cycle (ORC) module at design conditions.

Parameters	Unit	Full Electric Mode	CHP Mode
Input thermal power	kW	1100	
Inlet temperature of the heat transfer fluid (water)	°C	≥160	
Outlet temperature of the heat transfer fluid (water)	°C	140	
Heat transfer fluid (water) mass flow rate	kg/s	13.14	
Electric power	kW	150	98
Electric efficiency	%	13.6	8.6
Output thermal power	kW	940	1000
Inlet temperature of water for thermal use	°C	26	60
Outlet temperature of water for thermal use	°C	36	80
Water mass flow rate	kg/s	22.46	11.95

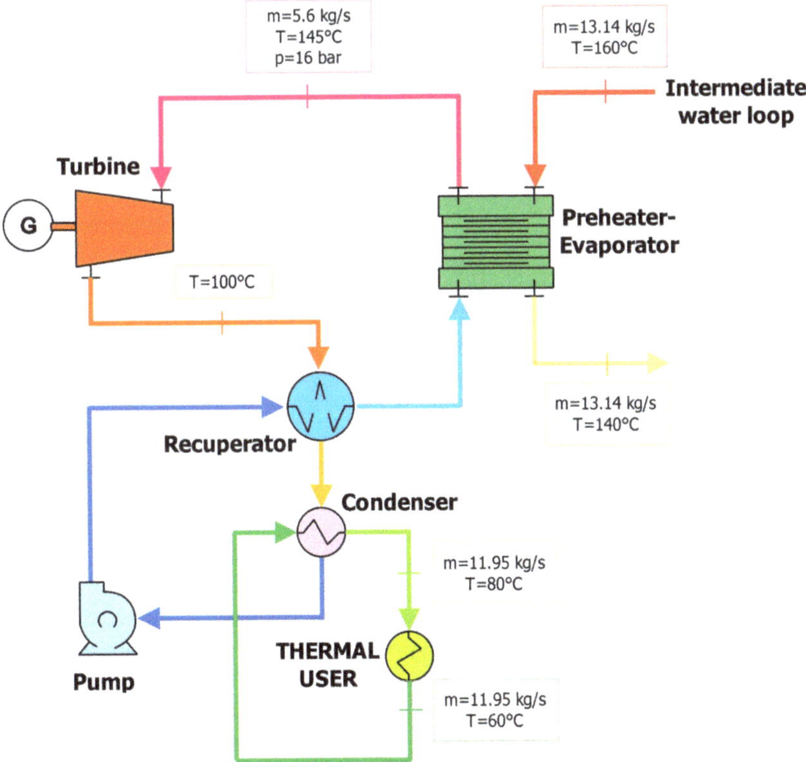

Figure 1. Schematic view of the ORC unit in its Combined Heat and Power (CHP) operation. Data refer to design point operating conditions.

Table 2. Devices of the ORC module.

Components	Characteristics
Pre-heater	Counter-flow brazed plate heat exchanger
Evaporator	Counter-flow brazed plate heat exchanger
Turbine	Radial turbine with fixed nozzles
Electric generator	Brushless synchronous generator with permanent magnets
Regenerative heat exchanger	Counter-flow brazed plate heat exchanger
Condenser	Counter-flow brazed plate heat exchanger
Pump	Centrifugal machine

To avoid risky contact between biomass combustion products and the ORC working fluid and to guarantee a steadier operation, the manufacturer adopts an intermediate water loop.

Thanks to this arrangement, biomass can be directly burnt in a commercial boiler.

The combustion products generated from biomass combustion heat up the pressurized water flowing in the intermediate loop, which in turn exchanges heat with the organic fluid.

The system includes also a cooling tower which allows to operate in full electric mode when users do not require heat.

In order to perform the Life Cycle Assessment analysis, data collected during five years of plant operation and from manufactures datasheets are organized in accordance with the provisions of ISO 14040 [38].

The system model is implemented in SimaPro environment [52].

SimaPro is one of the leading LCA software packages; it is a tool for the collection, analysis and monitoring of data on the environmental performance of goods and services. It makes available many internationally recognized sectoral databases, which allow the insertion of data and processes into inputs and outputs and includes the most recognized impact analysis methods.

In the system's model, data related to ORC devices are considered: weights, type of materials and mechanical processing.

Table 3 summarizes the materials used to build the ORC system. They comprise weights of pipes, valves, tank of the condensate, cooling tower, inverter, electrical cabinet and skid. For simplicity, materials are grouped for families, but in the inventory, the specific materials are considered. As an example, Table 3 lists steel, but in the inventory, carbon and stainless steel are separately accounted for.

Table 3. Summary of the mass streams of the ORC module.

Material	Unit	Weight
Steel	kg	5320
Brass	kg	25
Pig iron	kg	150
Copper	kg	510
Aluminium alloy	kg	2
Permanent magnets	kg	4.32

The processes of braze welding in the heat exchangers, the anti-rust treatment of the tank, the welding of the pipes and the heat galvanization treatment of the cooling tower are also included in the analysis as well as the transport of all the devices to the plant installation site.

As known, in ORC plant the working fluid is an organic compound. The under investigation ORC unit adopts as working fluid a mixture of hydrocarbons (previously named "HCF") and it requires about 362 kg of fluid.

From data collected during the five years of operation, it is possible to estimate both leakage of organic fluid in the environment and cooling tower refill water. Organic fluid leakage is expected to be about 2%/year, while the estimated cooling tower water refilling is approximately 26.4 kg/min.

Based on plant manufacturer experience in terms of both components lifespan and maintenance activities, a plant life equal to 15 years can be assumed.

Concerning the end of life scenario, it is expected that about 80% of the organic fluid is recovered at the end of plant life, while the remaining 20% is considered to be released in the environment [53].

Starting from the Italian recycling rates of steel and copper [54,55], it is reasonable to assume a recycling percentage for steel and copper equal to 70% and 90%, respectively. It is also assumed that non-recycled materials are sent to landfills.

3. Materials and Methods

The functional unit is 1 kWh of electricity production.

Two different operating conditions are investigated:

- Plant working in full electric mode for 6000 h/year. This operating condition corresponds to a production during the entire life of the plant of 13,500 MWh.
- Plant running in CHP mode for 3000 h/year while, for the remainder 3000 h/year, it works in full electric mode. Note that, this is a regular working condition for plants serving residential users, as in the under-investigation case. It corresponds to an electricity production during the entire life of the plant of 11,160 MWh, and a heat production of 33,750 MWh.

In both cases, the overall biomass consumption is 99,000 MWh.

These data are added to the model with those related to plant construction and end of life, and to those associated to biomass cultivation and transport discussed in the previous section.

Figure 2 depicts a simplified tree representation of the ORC system implemented model.

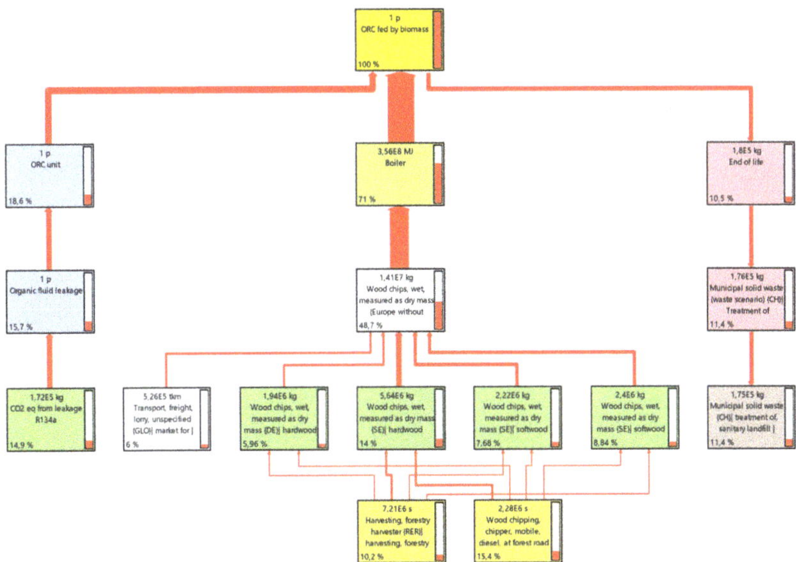

Figure 2. Simplified tree representation of the system in the lifecycle assessment. Percentages and lines refer to $CO_{2,eq}$ emissions.

As stated above, the commercial software SimaPro 9.0 [52] is used to build the ORC system model and to perform the Life Cycle Assessment analysis.

Among the different databases implemented in the software, Ecoinvent 3, Agri-footprint, USA Input–Output Database, EU and DK Input–Output Database, LCA Food DK are employed to study the processes of production, transport and disposal of each material constituting the plant. For electricity consumed during the construction and end of life phases, the Italian low voltage market mix is considered.

For impacts evaluation, the model implements four different methods:

- IPCC 2013 GWP100y [56], developed by the IPCC with a unique focus on the greenhouse effect. In this work, the analysis refers to the factors over 100 years, which quantify the medium-term effects of emissions.
- ReCiPe 2016, whose primary goal is transforming the long list of inventory results into a small number of significant indicators [57] but sufficient to describe the total effects of the process under examination on the environment. Here, the Hierarchical Perspective (H) is used, as suggested in [52], with "European" standardization and "medium" weighing set. Both Midpoint and Endpoint Indicators are computed.
- Cumulative Energy Demand, a method with a unique focus on the energy demand [58].
- Greenhouse Gas protocol [59], developed by the World Resources Institute and the World Business Council for Sustainable Development, with a unique focus on the carbon footprint. It considers also biogenic and untaken emissions.

4. Results and Discussion

Table 4 summarises the impacts computed with the ReCiPe 2016 method for the operation at full electric load, while Figure 3 depicts the contribution of the different operations on the overall impact.

Table 4. Summary of the mass stream of the ORC module.

Impact Category	Unit	Total
Global warming	kg $CO_{2,eq}$	0.085283
Stratospheric ozone depletion	kg $CFC11_{eq}$	5.53×10^{-7}
Ionizing radiation	kBq $Co\text{-}60_{eq}$	0.001301
Ozone formation. Human health	kg $NO_{x,eq}$	0.001811
Fine particulate matter formation	kg $PM2.5_{eq}$	0.000439
Ozone formation. Terrestrial ecosystems	kg $NO_{x,eq}$	0.00184
Terrestrial acidification	kg $SO_{2,eq}$	0.001447
Freshwater eutrophication	kg P_{eq}	5.83×10^{-6}
Marine eutrophication	kg N_{eq}	8.33×10^{-6}
Terrestrial ecotoxicity	kg 1.4-DCB	1.68345
Freshwater ecotoxicity	kg 1.4-DCB	0.007768
Marine ecotoxicity	kg 1.4-DCB	0.01133
Human carcinogenic toxicity	kg 1.4-DCB	0.002995
Human non-carcinogenic toxicity	kg 1.4-DCB	0.510964
Land use	m^2a $crop_{eq}$	0.87156
Mineral resource scarcity	kg Cu_{eq}	0.000275
Fossil resource scarcity	kg oil_{eq}	0.015262
Water consumption	m^3	0.009958

Focusing on global warming, total emissions of climate-changing gases are equal to 85.2 g $CO_{2,eq}$/kWh; a value definitively lower than that of the production from fossil fuels, which is estimated around 500 g $CO_{2,eq}$/kWh for the Italian fossil mix [60].

Note that the organic fluid leakages in the ORC affects this value for about 18.6%. Then, greater attention needs to be paid to this aspect because a better management of the leakages can remarkably improve the impact.

The recycling of steel and copper gives an appreciable positive impact on the mineral resource scarcity and Freshwater Eutrophication.

A large amount of water is required; but, for about 98% it is computable to the water evaporation in the cooling tower.

For many categories, biomass constitutes the major impact. This fact is related to the transportation of wood and tinder, while combustion is responsible for ODP and ecotoxicity high values due to the release of particulate emissions and ash.

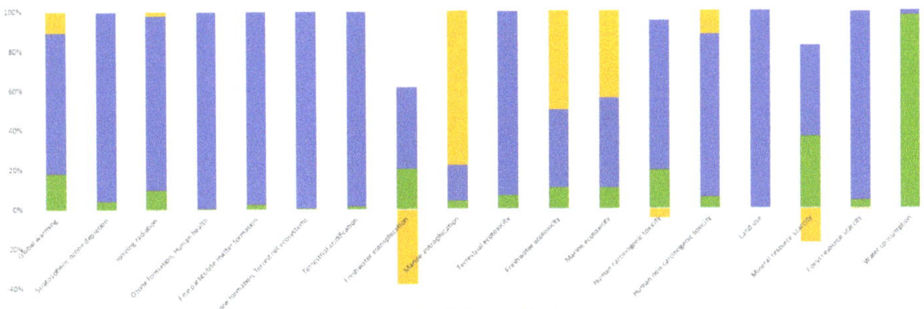

Figure 3. Contribution of operations in full electric mode (ReCiPe 2016 method).

Figure 4 summarizes the results of the ReCiPe Endpoint (H) method and shows the shares of different impacts on Human Health. It is clear that the emissions of particulate matter are the most impacting factor, followed by the non-carcinogenic and climate change emissions. All these emissions are mainly related to the combustion of biomass, and secondarily to the ash management.

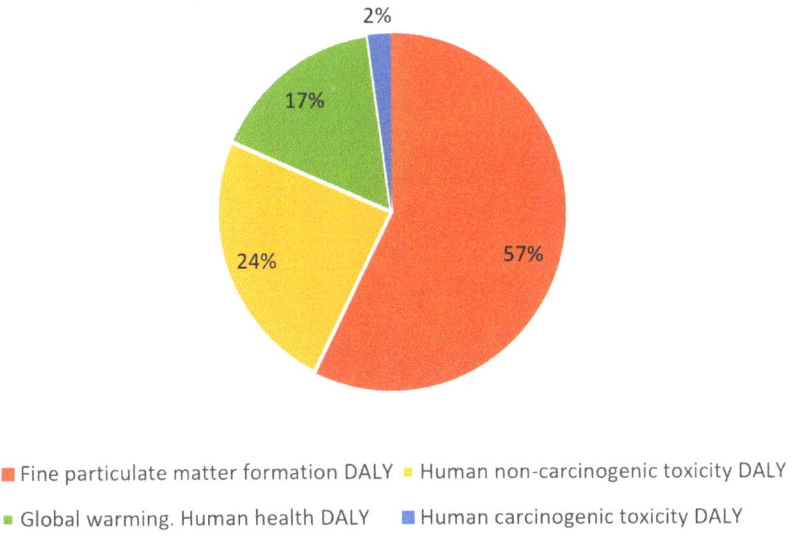

Figure 4. Contribution to Human Health of different impacts. Results are obtained with ReCiPe 2016—Endpoint (H) method.

Similar results are obtained in [61], where emissions related to the combustion and the transport process of the biomass have the largest share on overall plant impact.

The shares of different impacts on the ecosystem are reported in Figure 5. In this case, the land consumption linked to biomass cultivation is absolutely the most important factor.

In the case of managing the ORC in CHP mode, the avoided use of domestic boilers needs to be considered in the LCA analysis. Then, as shown in Figure 6, it is clear that managing the unit in CHP mode helps to reduce the overall impact. For many impact categories, the contribution of the ORC system becomes negative because the benefit introduced by the avoided use of natural gas in domestic boilers is higher than the impact generated by the ORC plant.

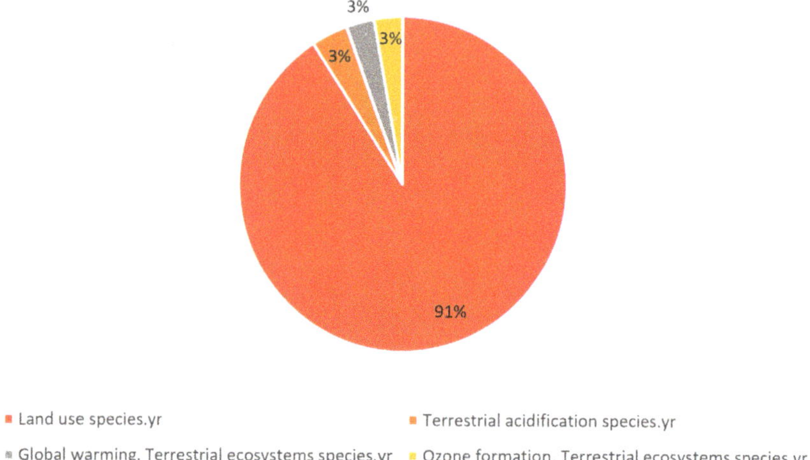

Figure 5. Contribution to Ecosystem of different impacts. Results are obtained with ReCiPe 2016—Endpoint (H) method.

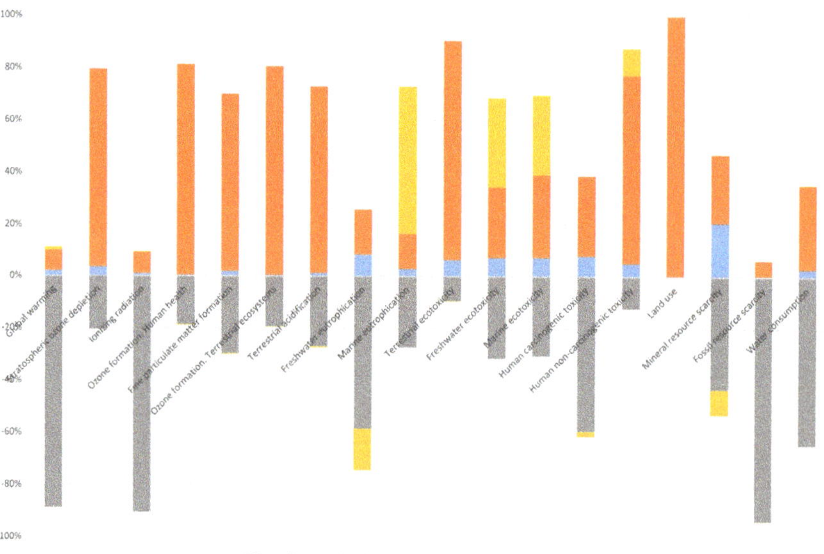

Figure 6. Contribution of the different operations to the overall impact in CHP mode. Results are obtained with ReCiPe 2016—Midpoint (H) method.

Figure 7 presents a focus on the ORC unit since it does not consider the biomass cultivation and combustion. In particular, the contribution of each device constituting the ORC unit is highlighted. It is clear that the need of an organic fluid is the source of high impact for two categories: Climate Change and Ozone Depletion, while the skid, the regenerator and the condenser exhibit a high-impact on a large number of categories due to the involvement of high mass of materials used to build these components.

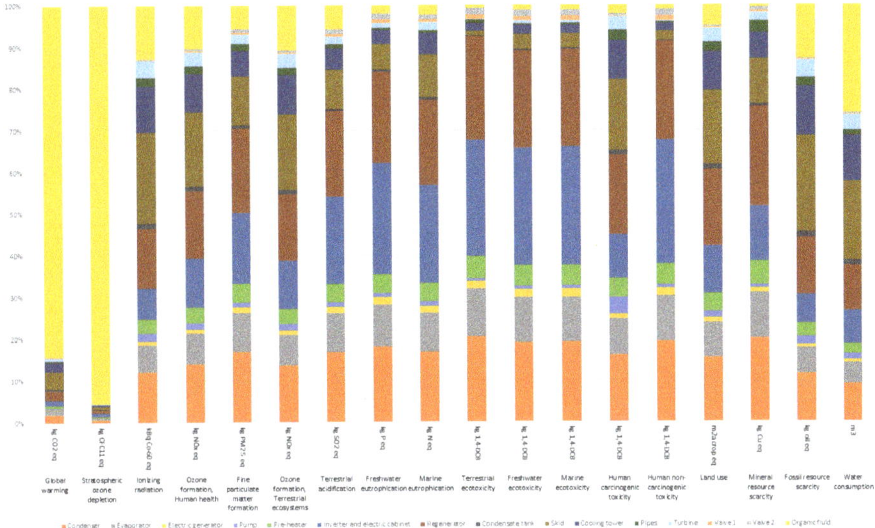

Figure 7. Contribution on the different impact categories of ORC devices construction and operation. Results are obtained with ReCiPe 2016—Midpoint (H) method.

Focusing on carbon footprint (see Figure 8 and Table 5), the impact of the combustion is clearly predominant as regards the biogenic carbon dioxide equivalent emitted, but biomass has a large impact also as regards to the emission of fossil carbon dioxide, which is of greater interest. Although the contribution of this phase is still prevalent, it accounts only for 79%; 21% of the emissions are instead determined by the production, assembly, disposal of the ORC module and by the discharge of part of the organic fluid into the atmosphere.

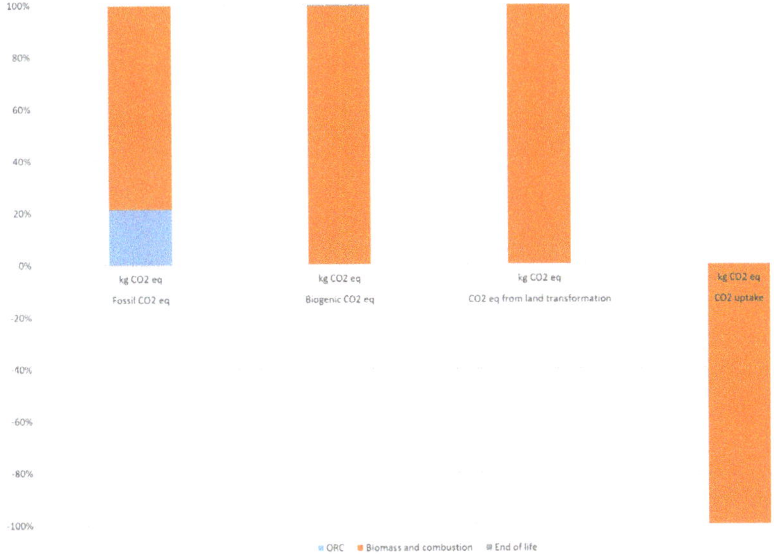

Figure 8. Carbon footprint.

Table 5. Summary of the carbon footprint.

Impact Category	g $CO_{2,eq}$
Fossil	73.2
Biogenic	1910
from land transformation	0.89
CO_2 uptake	1900

For a full grasp of these specific simulation outcomes, it is important to underline that "biomass and combustion" component includes, besides combustion, the phase of cultivation, processing and transport of biomass, boiler construction and use of pollutant abatement systems.

In order to evaluate the possibility of improving the life cycle performances of the system, as suggested by other researchers (see, e.g., [41]), the LCA analysis needs to be performed adopting also different organic fluids. In this analysis, R245fa and R1233zd are selected and their behavior compared to the "HCF" fluid originally used in the ORC unit.

Both fluids are characterized by an ODP equal to 0 as in the case of "HCF". The GWP amounts to 1030 in the case of R245fa (8.4% higher than the one of the actually adopted fluid) while it is equal to 1 for R1233zd (950 times lower than the "HCF" one). R1233zd is a low GWP fluid design to substitute R123 but it still constitutes a valid replacement to R245fa. A preliminary simulation of the ORC permits to evaluate the efficiency of the cycle with the new fluids and their required amounts.

Figure 9 summarizes the results obtained with the ReCiPe 2016 method. It is clear that for some categories, such as climate change and ecotoxicity, the impact using R1233zd is really reduced. Thus, this fluid guarantees the same energetic performance of the "HCF" but an absolutely lower impact on the entire set of impacts categories.

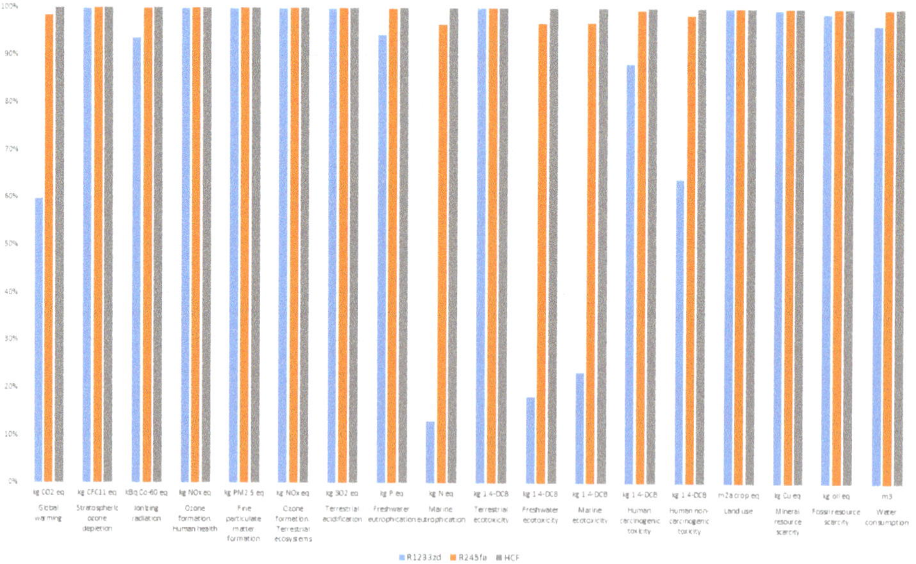

Figure 9. Comparisons of the life cycle impacts for three different organic fluids. Results are obtained with ReCiPe 2016—Midpoint (H) method.

These results are in agreement with Ref. [41], which studies a geothermal ORC plant. Obviously, the comparison between the authors and reference outcomes is not simple since a geothermal plant needs a lot of infrastructures, very impacting on the environment, which are not necessary when biomass is used. On the other hand, a geothermal unit does not require combustion and continuous biomass consumption which in turn, means continuous biomass cultivation and transportation to the site. In addition, the ORC lifetime estimation for the two plants is very different.

In any case, the improvement of the climate change impact obtained using R1233zd fluid instead of R245fa in the present analysis is about 60 g CO_2/kWh which is similar to the value obtained in Ref. [41].

In addition, also in Ref. [41] the analysis confirms that the other impact categories are marginally modified by the use of different fluids.

Lastly, the CED method shows that for each kWh of electricity, the system uses about 7.3 kWh of biomass, but also about 0.24 kWh of fossil fuels, mainly due to the need of Diesel for biomass transportation, chipping and harvesting.

5. Conclusions

Renewable energy sources and energy efficiency are considered the most useful ways to tackle climate change and shift towards a low-carbon economy.

To this purpose, economic supports have been established throughout the years and the main outcome is the widespread of renewable plants, while energy efficiency still lacking especially in the renewable energy sources field.

Among renewables, biomass and, in particular, solid biomass like woodchip, constitutes a valid option to generate electricity, heat or a combination of them due to high fuel supplies programmability and boilers efficiency.

In the energy efficiency sector, the organic Rankine cycle is a consolidate technology which guarantees good efficiency and high reliability.

Therefore, biomass boilers coupled with organic Rankine cycles are a viable option that guarantees to generate electricity and heat.

Despite biomass renewability and ORC unit capability of recovering the waste heat coming from biomass combustion products, it is mandatory to examine system environmental impact taking into account for the fuel supply chain and impacts on the environment caused by organic fluid losses linked to plant leakages.

To this purpose and by means of the lifecycle assessment method, the authors study the environmental impacts of a system composed by a biomass boiler and a 150 kW cogenerative ORC unit.

The analysis is not based on simulation results, but on data collected during five years of plant operation. This aspect is crucial when a life cycle approach is used.

Results show that the most impacting processes in terms of CO_2 equivalent emissions are the ones related to biomass production and organic fluid leakages. They accounted for 71% and 19% of the total, respectively.

Therefore, being fluid release into the environment high impacting, a comparison among three fluids are also performed and presented.

The analysis reveals that adopting R1233zd instead of the hydrocarbons fluid actually mounted in the cycle guarantees an improvement in all the impact categories but especially in the climate change one.

It is also finally important to remark that the CED method shows that for each kWh of electricity, the system composed by a biomass boiler and an ORC uses about 7.3 kWh of biomass and approximately 0.24 kWh of fossil fuels because there is the need of Diesel fuel for biomass transportation, chipping and harvesting.

In future works, the analysis will be extended to different in-operation plants based on the same commercial ORC unit. This is fundamental to evaluate how different management strategies, for example related to the transport and the cultivation of biomass or to CHP mode, can affect the overall impact of the plant.

Author Contributions: conceptualization, A.S. and A.B.; data curation, A.S. and A.B.; investigation, A.S. and A.B.; writing—original draft, A.B.; writing—review and editing, A.S. and A.B. All authors have read and agreed to the published version of the manuscript.

Funding: This research received no external funding.

Conflicts of Interest: Authors declare no conflict of interest.

Nomenclature

1.4-DCB	1.4-DICHLOROBENZENE
CFC11	Trichlorofluoromethane
CHP	Combined Heat and Power
CO_2	Carbon Dioxide
Cu	Copper
eq	equivalent
EU	European Union
g	gramme
GHG	Green House Gases
GT	Gas Turbine
GWP	Global Warming Potential
HCF	hydrocarbons fluid
ICE	Internal Combustion Engine
kBq Co-60eq	kilobecquerels of cobalt-60 equivalents
kg	kilogramme
LCA	Life Cycle Assessment
N	Nitrogen
NO_x	Nitrogen Oxide
ODP	Ozone Depletion Potential
ORC	Organic Rankine Cycle
P	Phosphorus
PM	Particulate Matter
RES	Renewable Energy Sources
SO_2	Sulphur Dioxide
SRC	Steam Rankine Cycle
WHRU	Waste Heat Recovery Unit

References

1. BP. *BP Statistical Review of World Energy 2018*; BP: London, UK, 2018.
2. International Energy Agency. *World Energy Outlook 2018 Summary*; Technical report; International Energy Agency: Paris, France, 2018. [CrossRef]
3. International Renewable Energy Agency (IRENA). *Global Energy Transformation: A Roadmap to 2050*; International Renewable Energy Agency: Abu Dhabi, UAE, 2018. [CrossRef]
4. European Parliament and Council of the European Union. *Directive 2009/28/EC of the European Parliament and of the Council of 23 April 2009 on the Promotion of the Use of Energy from Renewable sources and Amending and Subsequently Repealing Directives 2001/77/EC and 2003/30/EC*; European Commission: Brussels, Belgium, 2009.
5. European Council . *2030 Framework for Climate and Energy Policies*; European Commission: Brussels, Belgium, 2014.
6. European Council. *EU Roadmap 2050*; European Commission: Brussels, Belgium, 2010.
7. Stoppato, A.; Benato, A.; Mirandola, A. Assessment of stresses and residual life of plant components in view of life-time extension of power plants. In Proceedings of the 25th International Conference on Efficiency, Cost, Optimization, Simulations and Environmental Impact of Energy Systems (ECOS 2012), Perugia, Italy, 26–29 June 2012.
8. Benato, A.; Stoppato, A.; Bracco, S. Combined cycle power plants: A comparison between two different dynamic models to evaluate transient behaviour and residual life. *Energy Convers. Manag.* **2014**, *87*, 1269–1280. [CrossRef]

9. Benato, A.; Stoppato, A.; Mirandola, A. Dynamic behaviour analysis of a three pressure level heat recovery steam generator during transient operation. *Energy* **2015**, *90*, 1595–1605. [CrossRef]

10. Benato, A.; Stoppato, A. Pumped Thermal Electricity Storage: A technology overview. *Therm. Sci. Eng. Prog.* **2018**, *6*, 301–315. [CrossRef]

11. Benato, A.; Stoppato, A. Energy and cost analysis of an Air Cycle used as prime mover of a Thermal Electricity Storage. *J. Energy Storage* **2018**, *17*, 29–46. [CrossRef]

12. Benato, A.; Stoppato, A. Integrated Thermal Electricity Storage System: Energetic and cost performance. *Energy Convers. Manag.* **2019**, *197*, 111833. [CrossRef]

13. Benato, A.; Stoppato, A. Energy and Cost Analysis of a New Packed Bed Pumped Thermal Electricity Storage Unit. *J. Energy Resour. Technol. Trans. ASME* **2018**. [CrossRef]

14. Benato, A.; Stoppato, A. Heat transfer fluid and material selection for an innovative Pumped Thermal Electricity Storage system. *Energy* **2018**, *147*, 155–168. [CrossRef]

15. Park, H.; Baldick, R. Integration of compressed air energy storage systems co-located with wind resources in the ERCOT transmission system. *Int. J. Electr. Power Energy Syst.* **2017**, *90*, 181–189. [CrossRef]

16. Benato, A. Performance and cost evaluation of an innovative Pumped Thermal Electricity Storage power system. *Energy* **2017**, *138*, 419–436. [CrossRef]

17. Benato, A.; Stoppato, A.; Mirandola, A. State-of-the-art and future development of sensible heat thermal electricity storage systems. *Int. J. Heat Technol.* **2017**, *35*, S244–S251. [CrossRef]

18. WBA. *Global Bioenergy Statistics 2018*; Technical Report; WBA: West Bromwich, UK, 2018.

19. Pereira, R.G.; Oliveira, C.D.; Oliveira, J.L.; Oliveira, P.C.P.; Fellows, C.E.; Piamba, O.E. Exhaust emissions and electric energy generation in a stationary engine using blends of diesel and soybean biodiesel. *Renew. Energy* **2007**. [CrossRef]

20. Benato, A.; Macor, A. Italian Biogas Plants: Trend, Subsidies, Cost, Biogas Composition and Engine Emissions. *Energies* **2019**, *12*, 979. [CrossRef]

21. Chiaramonti, D.; Rizzo, A.M.; Spadi, A.; Prussi, M.; Riccio, G.; Martelli, F. Exhaust emissions from liquid fuel micro gas turbine fed with diesel oil, biodiesel and vegetable oil. *Appl. Energy* **2013**. [CrossRef]

22. Benato, A.; Macor, A.; Rossetti, A. Biogas Engine Emissions: Standards and On-Site Measurements. *Energy Procedia* **2017**, *126*, 398–405. [CrossRef]

23. Nikpey Somehsaraei, H.; Mansouri Majoumerd, M.; Breuhaus, P.; Assadi, M. Performance analysis of a biogas-fueled micro gas turbine using a validated thermodynamic model. *Appl. Therm. Eng.* **2014**. [CrossRef]

24. Benato, A.; Macor, A. Biogas Engine Waste Heat Recovery Using Organic Rankine Cycle. *Energies* **2017**, *10*, 327. [CrossRef]

25. Macor, A.; Benato, A. Experimental Measurements and Damage Assessment on Human Health. *Energies* **2020**, *13*, 1044. [CrossRef]

26. Vélez, F.; Segovia, J.J.; Martin, M.C.; Antolin, G.; Chejne, F.; Quijano, A. A technical, economical and market review of organic Rankine cycles for the conversion of low-grade heat for power generation. *Renew. Sustain. Energy Rev.* **2012**, *16*, 4175–4189. [CrossRef]

27. Benato, A.; Stoppato, A.; Mirandola, A. Renewable Energy Conversion and Waste Heat Recovery Using Organic Rankine Cycles. In *Renewable Energy Systems*; Kale, S.A., Ed.; Nova Science Publishers: Hauppauge, NY, USA, 2016; Chapter 11.

28. Bianchi, M.; De Pascale, A. Bottoming cycles for electric energy generation: Parametric investigation of available and innovative solutions for the exploitation of low and medium temperature heat sources. *Appl. Energy* **2011**, *88*, 1500–1509. [CrossRef]

29. Pezzuolo, A.; Benato, A.; Stoppato, A.; Mirandola, A. The ORC-PD: A versatile tool for fluid selection and Organic Rankine Cycle unit design. *Energy* **2016**, *102*, 605–620. [CrossRef]

30. Benato, A. Improving the efficiency of a cataphoresis oven with a cogenerative organic Rankine cycle unit. *Therm. Sci. Eng. Prog.* **2018**, *5*, 182–194. [CrossRef]

31. Pezzuolo, A.; Benato, A.; Stoppato, A.; Mirandola, A. Recovering gas turbine high-temperature exhaust heat using organic Rankine cycle with mixture as working fluid. In Proceedings of the 29th International Conference on Efficiency, Cost, Optimization, Simulations and Environmental Impact of Energy Systems (ECOS 2016), Portoroz, Slovenia 19–26 June 2016.

32. Campana, F.; Bianchi, M.; Branchini, L.; De Pascale, A.; Petto, A.; Baresi, M.; Fermi, A.; Rossetti, N.; Vescovo, R. ORC waste heat recovery in European energy intensive industries: Energy and GHG savings. *Energy Convers. Manag.* **2013**, *76*, 244–252. [CrossRef]

33. Branchini, L.; De Pascale, A.; Peretto, A. Systematic comparison of {ORC} configurations by means of comprehensive performance indexes. *Appl. Therm. Eng.* **2013**, *61*, 129–140. [CrossRef]

34. Cavazzini, G.; Bari, S.; Pavesi, G.; Ardizzon, G. A multi-fluid PSO-based algorithm for the search of the best performance of sub-critical Organic Rankine Cycles. *Energy* **2017**. [CrossRef]

35. Benato, A.; Cavazzini, G.; Bari, S.; Ardizzon, G. ORC pump efficiency estimation and real behaviour under different working fluids. In Proceedings of the 32nd International Conference on Efficiency, Cost, Optimization, Simulation and Environmental Impact of Energy Systems (ECOS 2019), Wrocław, Poland, 23–28 June 2019.

36. Ancona, M.A.; Bianchi, M.; Branchini, L.; De Pascale, A.; Melino, F.; Orlandini, V.; Ottaviano, S.; Peretto, A.; Pinelli, M.; Spina, P.R.; et al. A Micro-ORC Energy System: Preliminary Performance and Test Bench Development. *Energy Procedia* **2016**. [CrossRef]

37. Ancona, M.A.; Bianchi, M.; Branchini, L.; Pascale, A.; Melino, F.; Ottaviano, S.; Peretto, A.; Scarponi, L.B. Heat recovery from a liquefied natural gas production process by means of an organic rankine cycle. In Proceedings of the ASME Turbo Expo, Lillestrøm, Norway, 11 June 2018. [CrossRef]

38. ISO. *14040: Environmental Management–Life Cycle Assessment—Principles and Framework*; International Organization for Standardization: Geneva, Switzerland, 2006.

39. Stougie, L.; Tsalidis, G.A.; van der Kooi, H.J.; Korevaar, G. Environmental and exergetic sustainability assessment of power generation from biomass. *Renew. Energy* **2018**. [CrossRef]

40. Walsh, C.; Thornley, P. The environmental impact and economic feasibility of introducing an organic Rankine cycle to recover low grade heat during the production of metallurgical coke. *J. Clean. Prod.* **2012**. [CrossRef]

41. Liu, C.; He, C.; Gao, H.; Xie, H.; Li, Y.; Wu, S.; Xu, J. The environmental impact of organic Rankine cycle for waste heat recovery through life-cycle assessment. *Energy* **2013**. [CrossRef]

42. Heberle, F.; Schifflechner, C.; Brüggemann, D. Life cycle assessment of Organic Rankine Cycles for geothermal power generation considering low-GWP working fluids. *Geothermics* **2016**. [CrossRef]

43. Lin, Y.P.; Wang, W.H.; Pan, S.Y.; Ho, C.C.; Hou, C.J.; Chiang, P.C. Environmental impacts and benefits of organic Rankine cycle power generation technology and wood pellet fuel exemplified by electric arc furnace steel industry. *Appl. Energy* **2016**. [CrossRef]

44. Uusitalo, A.; Uusitalo, V.; Grönman, A.; Luoranen, M.; Jaatinen-Värri, A. Greenhouse gas reduction potential by producing electricity from biogas engine waste heat using organic Rankine cycle. *J. Clean. Prod.* **2016**. [CrossRef]

45. Sedpho, S.; Sampattagul, S.; Chaiyat, N.; Gheewala, S.H. Conventional and exergetic life cycle assessment of organic rankine cycle implementation to municipal waste management: The case study of Mae Hong Son (Thailand). *Int. J. Life Cycle Assess.* **2017**. [CrossRef]

46. Monteleone, B.; Chiesa, M.; Marzuoli, R.; Verma, V.K.; Schwarz, M.; Carlon, E.; Schmidl, C.; Ballarin Denti, A. Life cycle analysis of small scale pellet boilers characterized by high efficiency and low emissions. *Appl. Energy* **2015**. [CrossRef]

47. Singh, P.; Gundimeda, H.; Stucki, M. Environmental footprint of cooking fuels: A life cycle assessment of ten fuel sources used in Indian households. *Int. J. Life Cycle Assess.* **2014**. [CrossRef]

48. Boschiero, M.; Kelderer, M.; Schmitt, A.O.; Andreotti, C.; Zerbe, S. Influence of agricultural residues interpretation and allocation procedures on the environmental performance of bioelectricity production—A case study on woodchips from apple orchards. *Appl. Energy* **2015**. [CrossRef]

49. Stoppato, A. Life cycle assessment of photovoltaic electricity generation. *Energy* **2008**. [CrossRef]

50. Fantinato, D.; Stoppato, A.; Benato, A. LCA analysis of a low-energy residential building. In Proceedings of the 32nd International Conference on Efficiency, Cost, Optimization, Simulation and Environmental Impact of Energy Systems (ECOS 2019), Wrocław, Poland, 23–28 June 2019; pp. 3153–3165.

51. Toscani, A.; Stoppato, A.; Benato, A. LCA of a concert: Evaluation of the Carbon footprint and of Cumulative energy demand. In Proceedings of the 32nd International Conference on Efficiency, Cost, Optimization, Simulation and Environmental Impact of Energy Systems (ECOS 2019), Wrocław, Poland, 23–28 June 2019; pp. 3203–3213.

52. Préconsultants. SimaPro 9.0. Available online: https://simapro.com/ (accessed on 8 April 2020).

53. Laleman, R.; Albrecht, J.; Dewulf, J. Life cycle analysis to estimate the environmental impact of residential photovoltaic systems in regions with a low solar irradiation. *Renew. Sustain. Energy Rev.* **2011**, *15*, 267–281. [CrossRef]

54. Hung, T.C.; Wang, S.K.; Kuo, C.H.; Pei, B.S.; Tsai, K.F. A study of organic working fluids on system efficiency of an ORC using low-grade energy sources. *Energy* **2010**. [CrossRef]

55. Qiu, G.; Shao, Y.; Li, J.; Liu, H.; Riffat, S.B. Experimental investigation of a biomass-fired {ORC}-based micro-{CHP} for domestic applications. *Fuel* **2012**, *96*, 374–382. [CrossRef]

56. IPCC. *Revised Supplementary Methods and Good Practice Guidance Arising from the Kyoto Protocol, Intergovernmental Panel on Climate Change*; Technical Report; IPCC: Geneva, Switzerland, 2013.

57. Huijbregts, M.; Steinmann, Z.J.N.; Elshout, P.M.F.M.; Stam, G.; Verones, F.; Vieira, M.D.M.; Zijp, M.; van Zelm, R. *ReCiPe 2016*; National Institute for Public Health and the Environment: Bilthoven, The Netherlands, 2016. [CrossRef]

58. Frischknecht, R.; Wyss, F.; Büsser Knöpfel, S.; Lützkendorf, T.; Balouktsi, M. Cumulative energy demand in LCA: the energy harvested approach. *Int. J. Life Cycle Assess.* **2015**. [CrossRef]

59. Schmitz, S.; Dawson, B.; Spannagle, M.; Thomson, F.; Koch, J.; Eaton, R. *The Greenhouse Gas Protocol—A Corporate Accounting and Reporting Standard*; Revised ed.; The GHG Protocol Corporate Accounting and Reporting Standard; World Resources Institute and World Business Council for Sustainable Development: Washington, DC, USA, 2001.

60. Istituto Superiore per la Protezione e la Ricerca Ambientale. *Fattori di emissione per la produzione ed il consumo di energia elettrica in Italia (Emission Factors for Production and Consumption of electricity)*; Technical Report; 2019. Available online: http://www.sinanet.isprambiente.it/it/sia-ispra/serie-storiche-emissioni/ (accessed on 8 April 2020).

61. Stougie, L.; Giustozzi, N.; van der Kooi, H.; Stoppato, A. Environmental, economic and exergetic sustainability assessment of power generation from fossil and renewable energy sources. *Int. J. Energy Res.* **2018**. [CrossRef]

Article

Thermodynamic, Exergy and Environmental Impact Assessment of S-CO$_2$ Brayton Cycle Coupled with ORC as Bottoming Cycle

Edwin Espinel Blanco [1], Guillermo Valencia Ochoa [2,*] and Jorge Duarte Forero [2]

[1] Facultad de Ingeniería, Universidad Francisco de Paula Santander, Vía Acolsure. Sede el Algodonal Ocaña, Ocaña-Norte de Santander 546552, Colombia; eeespinelb@ufps.edu.co

[2] Programa de Ingeniería Mecánica, Universidad del Atlántico, Carrera 30 Número 8–49, Puerto Colombia, Barranquilla 080007, Colombia; jorgeduarte@mail.uniatlantico.edu.co

* Correspondence: guillermoevalencia@mail.uniatlantico.edu.co; Tel.: +575-324-94-31

Received: 22 March 2020; Accepted: 29 April 2020; Published: 4 May 2020

Abstract: In this article, a thermodynamic, exergy, and environmental impact assessment was carried out on a Brayton S-CO$_2$ cycle coupled with an organic Rankine cycle (ORC) as a bottoming cycle to evaluate performance parameters and potential environmental impacts of the combined system. The performance variables studied were the net power, thermal and exergetic efficiency, and the brake-specific fuel consumption (BSFC) as a function of the variation in turbine inlet temperature (TIT) and high pressure (P$_{HIGH}$), which are relevant operation parameters from the Brayton S-CO$_2$ cycle. The results showed that the main turbine (T1) and secondary turbine (T2) of the Brayton S-CO$_2$ cycle presented higher exergetic efficiencies (97%), and a better thermal and exergetic behavior compared to the other components of the System. Concerning exergy destruction, it was found that the heat exchangers of the system presented the highest exergy destruction as a consequence of the large mean temperature difference between the carbon dioxide, thermal oil, and organic fluid, and thus this equipment presents the greatest heat transfer irreversibilities of the system. Also, through the Life Cycle Analysis, the potential environmental impact of the system was evaluated to propose a thermal design according to the sustainable development goals. Therefore, it was obtained that T1 was the component with a more significant environmental impact, with a maximum value of 4416 Pts when copper is selected as the equipment material.

Keywords: Brayton; environmental impact; exergy; life cycle analysis; ORC; performance parameters

1. Introduction

With the reinforcement of environmental legislation, emission mitigation continues to be an essential problem at the industrial level, and heat recovery in manufacturing procedures is becoming standard practice because of progressively stringent policies on energy efficiency [1]. Therefore, waste heat recovery systems play an essential role in saving energy by considering existing energy generation technologies that are geared towards reducing fuel consumption, greenhouse emissions, and electricity production cost [1]. For this reason, the volume of power and temperature, as well as the form and the prices of technologies for the recovering of waste resources, are the key factors determining the feasibility of energy consumed. Accordingly, to reach a maximum recovery capacity, it is of particular significance that the temperature and the residual heat correlate with the energy recovery periods for the remaining electricity [1].

For alternative energy generation, the concept of the simple regeneration process has been suggested, considering CO$_2$ as a working solvent due to its properties [2], including low critical strain, high thermal tolerance at interest temperature, inertia, well known thermal properties, as well as being nontoxic and economical. Angelino et al. [3] published research related to various cycle designs. They demonstrated that the recompression cycle is better at high temperatures, and it is particularly interesting in high-temperature gas-cooled reactors. Subsequently, Dostal et al. [4] in his thesis evaluated the Brayton S-CO_2 cycle for advanced nuclear power generation reactors.

Nowadays, many organic Rankine cycle (ORC) applications have been used as a waste heat recovery system to convert waste heat into mechanical energy. However, the ORC has efficiency limitations when working with waste heat at high temperatures because of the physical and thermal properties usually presented by organic fluids. Abrosimov et al. [5] investigated the combination of a Brayton cycle and an ORC cycle by designing both ORC and combined cycle models using Aspen Hysys® version 9' (Aspen Technology, Inc., Bedford, MA, USA. Thermodynamic and economic optimizations of the models were made to conduct a comparative analysis between the solutions. The results have demonstrated the 10% advantage of the combined scheme over the ORC cycle in terms of generated power and system efficiency. Optimization based on the leveled energy cost for variable capacity factors has revealed an advantage of more than 6% of the solution investigated.

Zhangpeng Guo et al. [6] conducted a sensitivity analysis comparing the recompression cycle, the double-expansion recompression cycle, and the modified recompression cycle applied to fourth-generation nuclear reactors, which have high operating temperatures and pressures that would increase plant efficiency and hydrogen production. Vasquez Padilla et al. [7] conducted a detailed energy and exergy analysis of four Brayton S-CO_2 cycle configurations (single Brayton cycle, recompression Brayton cycle, partial cooling recompression, and main compression with intercooling) with and without reheating to investigate the effect of replacing the reheater and heater by a solar receiver. In the same year, Ricardo Vásquez et al. [8] also performed the energy and exergy analysis of a supercritical Brayton cycle with recompression CO_2, but a bottoming cycle was not proposed.

In recent decades, the Brayton S-CO_2 cycle has attracted the attention of many academics and industries because of its significant advantages [3,9], such as a better thermal efficiency [10]. Brayton S-CO_2 is less caustic relative to steam with the same operating speed, the turbomachine used is lightweight, almost ten times smaller than the steam turbine, and dry cooled easily in comparison with the steam engine [11]. Therefore, based on each of these benefits, the Brayton S-CO_2 cycle has been tested for various uses as an energy conversion device, including nuclear, geothermal, solar, and thermal power plants. So, the device presents higher thermal performance and a smaller process size than the traditional Rankine steam device based on these advantages, and therefore the Brayton S-CO_2 cycle is considered an efficient alternative to the steam process at Rankine [12].

The thermal performance of the heat exchanger plays an important role in the cycle efficiency, as shown in these previously described studies [13]. Thus, when a significant amount of heat is extracted in the recuperator to improve thermal performance, the high energy output is expected and thus the capital cost decreases by utilizing traditional shell and tube heat exchangers (STHE); nonetheless, some high-compact heat exchangers (up to ten times relative to STHE) and printed circuit heat exchangers (PCHE) have been sold and can be added directly to waste heat recovery from gas [14]. However, this type of heat exchanger can be multi-objective optimized, attending to exergy and energy objective function [15]. Yuan Jiang et al. [16] published a paper detailing the core architecture and optimization methodology built into the Aspen Custom Modeler for microtube shell and tube exchangers. They also sized a PCHE and then compared it among the various heat exchangers used for S-CO_2 and indicated that the PCHE is a promising candidate due to its concentration, quick dynamic response, and mature construction state. Optimum configuration findings suggest that there is less metal mass in a system of two hot plates per cold plate and high angle channels; therefore, a safer option for broad-scale applications [17].

The PCHE is a type of micro-channel heat exchanger residing in various carve sheets composed of many micro-waved channels in each layer [18]. Thus, Devesh Ranjan et al. [19] in 2019 developed experimental research in which the characteristics of heat flow and pressure drop decrease for the

PCHE with a different configuration, which is a promising result increasing the thermal performance of the S-CO$_2$ integrated with ORC.

All this research involves a substantial economic investment; however, there is an effective way with a high degree of reliability to perform these tests without the need for tuning, and that is through dynamic modeling. In engineering, dynamic modeling and simulation are increasingly relevant, as there is a growing need to study the chaotic function of complex structures made up of components from different domains. The models may be used to conduct danger and operability tests. Automated emergency response protocols were validated in [20], as in the case of Xiaoyan Ji et al. [18] also, who performed the heat exchanger design for the suspensions in biogas plants, which was analyzed numerically based on the rheological properties and further coupled with full thermal cycles to demonstrate waste heat recovery and high heating capacity for heat exchangers design. Dynamic models may help design heat-sharing facilities [21]. For example, Oh Jong-Taek et al. [22] demonstrated employing computational fluid dynamics to study the heat transfer and flow characteristics, as well as the effect of mass flow on temperature and pressure distribution in PCHE.

On the other hand, the Brayton S-CO$_2$ cycles help to predict the advantages of compact equipment within the moderate temperature range (450–750 °C), as well as finding the disadvantages of materials due to high temperatures and pressure, studied in various applications [23,24]. Craig S. Turchi et al. [25] using simulations of the Brayton S-CO$_2$ cycle, observed favorable characteristics such as the capacity to adapt dry cooling and produce the desired efficiency in the area of solar energy concentration. Yann Le Moullec [26] proposed a closed Brayton S-CO$_2$ loop based on combining carbon capture and storage facilities to mitigate CO$_2$ pollution as electricity production from coal-fired power plants is a significant source of ambient CO$_2$ pollution, as Olumide Olumayegun et al. [27] where the thermodynamic performance of Brayton S-CO$_2$ cycles coupled to a coal furnace and integrated with 90% post-combustion CO$_2$ capture was evaluated, investigating three background s-CO$_2$ cycle designs, including a new recompression cycle with a single recuperator, showing as a result, that the configuration with a recompression cycle and a single recuperator has the highest net plant efficiency. Without CO$_2$ capture, the effectiveness of the coal-fired plants was higher than that of steam. Youcan Liang et al. [28], shows the application of the method on a dual-fuel engine, which reveals that the maximum net power of the system is up to 40.88 kW, improving by 6.78%, leading to greater energy efficiency and reduced fuel consumption of the engine. Complementarily, Shih-Ping Kao et al. [29] released the results of a complex simulation code based on actual gas and integral momentum models, which was developed at the Massachusetts Institute of Technology to test control strategies for a small light-water reactor fitted with a compact Brayton S-CO$_2$ cycle.

Based on the significant energy loss presented in the industrial generation engine, some waste heat recovery systems based on the Brayton S-CO$_2$ cycle have been studied. Therefore, in 2018 Antti Uusitalo et al. [30] examined the use of supercritical Brayton cycles to recover power from the exhaust gases of large-scale engines. The objective of this study was to examine electricity generation through varying operating conditions and organic fluids and thus define the key design parameters influencing the cycle's energy output. In 2018 Piero Danieli et al. [31] measured the economic and thermodynamic efficiency of four separate waste heat recovery systems implemented through simulation to two hollow glass furnaces producing around 4 MWt of heat loss at 450 °C. Also, Subhash Lahane et al. [32] planned a heat exchanger design to collect the excess heat from the exhaust gases of a diesel engine to preheat the air entering the combustion chamber; this should be situated between the engine's inlet and outlet ducts.

However, it is also possible to determine optimal operating conditions of the combined Brayton-ORC system through advanced exergetic analysis, as in the case of Valencia et al. [33], where the fluid is selected to perform an advanced exergetic analysis in a combined thermal system using the ORC as bottoming cycle of an internal combustion engine, finding improvements of up to 80% in the components of the process. The thermo-economic analyses had been used to optimize the components of a trigeneration system, constituted by a gas microturbine and a heat recovery steam generation sub-system [34]. In addition, the ORC as a bottoming cycle of an internal combustion engine was proposed, identifying the heat transfer equipment with the highest exergy destruction costs, representing 81.25% of the total system cost, and acetone as the working fluid with the best impact on reducing these costs [35]. Another approach to determining optimal operating conditions is to conduct mathematical modeling of the physical phenomena. So, a phenomenologically based semi-physical model for a 2 MW internal combustion engine was obtained as a function of average thermodynamic value, validated by real operating data, to predict the thermodynamic performance parameters of the bottoming cycle as waste heat recovery systems [36].

The above studies propose heat recovery with bottoming cycles, using the different thermal power configurations, modeling the cycle components or making simulations of the same, including economic optimization in one of them; leaving with a wide uncertainty gap regarding the performance of this cycle integrated to ORC as bottoming cycle, which, despite not handling high temperatures, is a challenge in terms of designing a highly efficient heat exchanger so that high thermal performance of the combined power cycle can be obtained.

Therefore, the main contribution of this research is to present a thermodynamic, exergy, and environmental parametric study of a proposed combined S-CO_2 and ORC as a bottoming cycle considering the toluene, cyclohexane, and acetone as the organic working fluids. The methodology of the life cycle analysis is applied to investigate the environmental impact of the components of the system during its lifetime, which allows for the evaluation of environmental impacts, and the determination of energetic and exergetic improvement potentials for the environmental sustainability of the system.

2. Methodology

2.1. Description and Properties of the System

The system shown in Figure 1 is an ORC configuration combined with a Brayton S-CO_2 cycle. The Brayton cycle consists of the following components: the primary turbine (T1), a secondary turbine (T2), an axial compressor (C1), a reheater (RH), and a recuperator (HTR). The ORC cycle includes a shell and tube heat exchanger (ITC1), an evaporator (ITC2), a condenser (ITC3), a thermal oil circuit pump (P1), an organic fluid pump (P2), and a turbine (T1).

In the process, the carbon dioxide enters the primary turbine (T1) of the Brayton cycle at point 1, at a high temperature and pressure, then enters to the reheater (RH) at point 2 and is expanded to lower pressure in the secondary turbine (T2). Next, by means of the recuperator (HTR), the carbon dioxide (point 7) that leaves the compressor (C1) is reheated, which is conducted to the heater (RH) to obtain the thermodynamic state reported as (point 8). Meanwhile, the fluid in point 5 is cooled by yielding heat to the thermal oil, and then it is compressed by the compressor (C1) at point 6.

In the ORC, the thermal oil (Therminol 75) receives heat in the heat exchanger (ITC1) to be transferred to the evaporator (ITC2). During this process, there are three stages: preheating, evaporation, and overheating, with the purpose of heat the organic fluid (Cyclohexane, Toluene, Acetone), while the thermal oil flow through the cycle with the energy supply by the pump (P1). Then, the organic fluid enters the turbine (T3) at a high temperature and pressure through the 1ORC current and expands to decrease its pressure and temperature to enter the condenser (ITC3), where the water cools it at room temperature (point 1A), and then goes to the reservoir (2A). Subsequently, when the organic working fluid leaves the condenser (ITC3) at point 3ORC, it enters the pump (P2) as a saturated liquid and then completes the cycle entering the evaporator (ITC2).

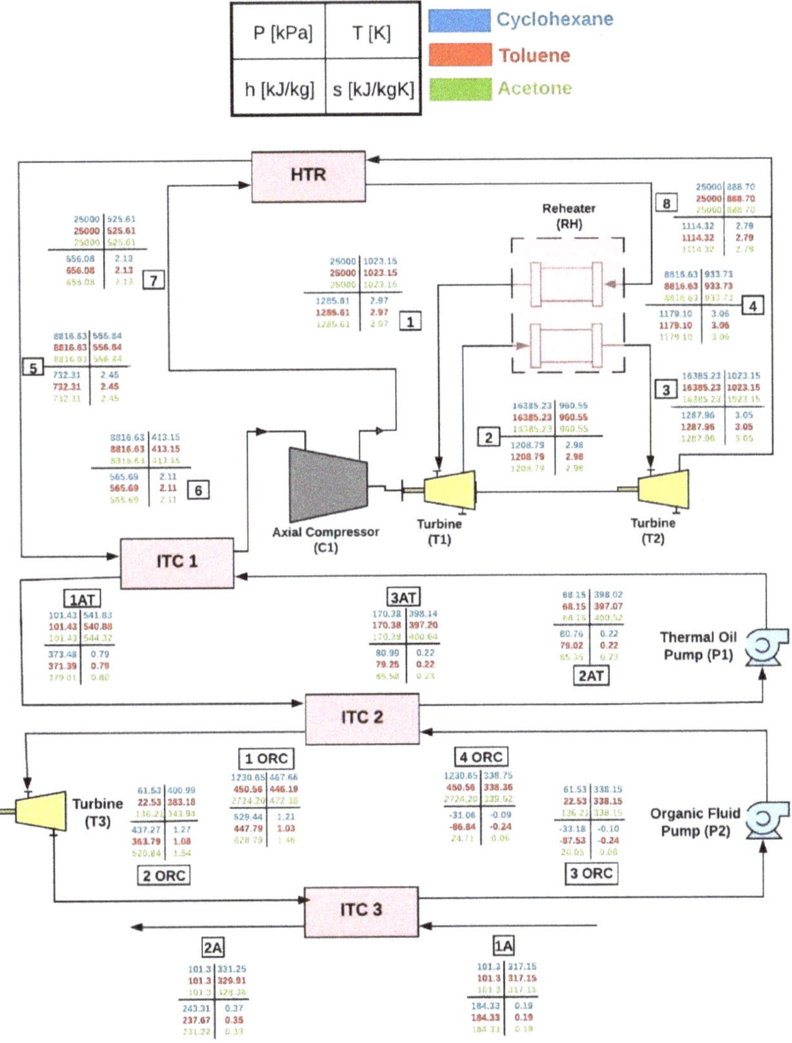

Figure 1. Physical structure of the Brayton S-CO$_2$ integrated into an organic Rankine cycle (ORC) as a bottoming cycle.

2.2. Working Fluids Properties

To determine the thermodynamic properties in the ORC cycle, the types of fluids used in the system are determined through the T-s diagram (Figure 2). The working fluids are categorized in this diagram according to the slope of the saturation vapor line, and it is known that the dry fluid displays a positive slope in the diagram, a negative slope in the wet fluids, and an extremely broad slope in the isentropic fluids.

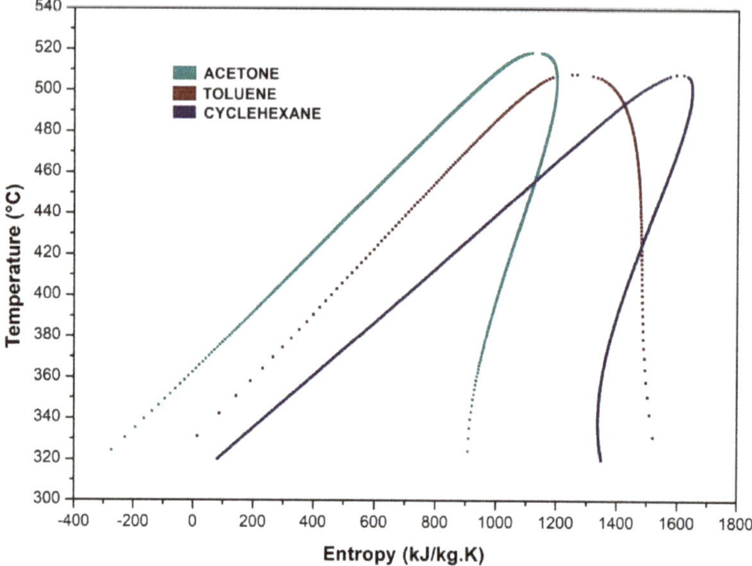

Figure 2. Entropy Temperature (T-s) diagram of organic fluids.

According to Equation (1), the fluid type is defined based on the slope of the saturation line. This slope classifies the working fluids by value of $E = \frac{\partial s}{\partial T}$ obtaining for dry fluids E > 0 isentropic fluids, E ≈ 0, and wet fluids E < 0 [37].

$$E = \frac{C_p}{T_H} - \frac{n \cdot T_H - T_{rH} + 1}{T_H^2 \cdot (1 - T_{rH})} \Delta H_H \tag{1}$$

where $T_{rH} = \frac{T_H}{T_C}$, and ΔH_H is the enthalpy of evaporation.

2.3. Thermodynamic Analysis

The consideration adopted for carrying out the combined Brayton S-CO$_2$ and ORC thermodynamic modeling are listed below [29]:

- The components were assumed as open systems in a stable state condition.
- The atmospheric temperature and pressure were set to 25 °C and 101.3 kPa, respectively.
- The kinetic and potential energy was not considered in the energy balance.
- The Brayton-CO$_2$ compressor is studied under similar turbine conditions.
- The consequences of pressure fluctuations are not known.
- Pressure drops in the tubing are not known.
- Pressure drops in structure heat exchangers are estimated depending on the geometry and flow regime characteristics.

The energy balance in steady-state for the devices is presented in Equation (2)

$$\dot{Q} - \dot{W} + \sum \dot{m}_{in} \cdot h_{in} - \sum \dot{m}_{out} \cdot h_{out} = 0 \tag{2}$$

where $\dot{Q}$ is the heat at the boundary in kW, $\dot{W}$ is the conversion of energy by work in kW, h is the specific enthalpy value for the working fluid in kJ/kg·K, and $\dot{m}$ is the mass flow rate in kg/s.

To calculate the working fluid exergy as a function of the ambient conditions, Equation (3) is used.

$$e_i = h_i - h_0 + T_0 s_0 - T_0 s_i \tag{3}$$

For each device of the cycle, the exergy destruction can be determined with the exergy balance described in Equation (4)

$$\dot{ED}_{di} = \dot{ED}_{Q_i} - \dot{ED}_{W_i} + \sum \dot{m}_{in} \cdot e_{in} - \sum \dot{m}_{out} \cdot e_{out} \tag{4}$$

where $\dot{ED}_{di}$ is the exergy destruction rate for components (i), the exergy rate by work and heat movement over the boundary is $\dot{ED}x_{W_i}$ and $\dot{ED}x_{Q_i}$, and the inlet and outlet related exergy rates are ex_{in} and ex_{out}.

The exergy rate by heat transfer is determined with Equation (5)

$$\dot{ED}x_{Q_i} = \dot{Q}_i \cdot \left(1 - \frac{T_0}{T_s}\right) \tag{5}$$

where T_0 is the atmospheric temperature, and T_s is the source temperature if the heat is produced and the temperature decreases when the heat is lost in the system. Also, Equation (6) is often used to quantify the exergy destruction rate by component ($\dot{ED}_{di}$)

$$\dot{ED}_{di} = T_0 \cdot \dot{s}_{gen.i} \tag{6}$$

where $\dot{s}_{gen.i}$ is the rate of entropy production, which is calculated with the general entropy balance with Equation (7), as shown as follows:

$$\dot{s}_{gen.i} = \sum \dot{m}_{out} \cdot s_{out} - \sum \dot{m}_{in} \cdot s_{in} - \sum \frac{\dot{Q}}{T} \tag{7}$$

The net power of the Brayton cycle ($\dot{W}_{net,Brayton\ S-CO_2}$) is calculated based on Equation (8), from the power of the main turbine (T1), the secondary turbine (T2) and the compressor (C1).

$$\dot{W}_{net,\ Brayton\ S-CO_2} = \dot{W}_{T1} + \dot{W}_{T2} - \dot{W}_{C1} \tag{8}$$

Equation (9) determines the net power of the ORC cycle ($\dot{W}_{net,ORC}$), based on the power of the turbine (T3) and pumps (P1 and P2).

$$\dot{W}_{net,ORC} = \dot{W}_{T3} - \dot{W}_{P1} - \dot{W}_{P2} \tag{9}$$

The ORC thermal efficiency ($\eta_{I,ORC}$) can be written based on Equation (10).

$$\eta_{I,ORC} = \frac{\dot{W}_{net,ORC}}{\dot{Q}_{ITC1}} \tag{10}$$

where $\dot{W}_{net,ORC}$ is the net power of the ORC, and $\dot{Q}_{ITC1}$ is the heat collected from the heat exchanger.

Equation (11) is used to calculate the thermal performance of the Brayton cycle as a function of the net power of the Brayton cycle and the heat received from the thermal source ($\dot{Q}_{RH}$).

$$\eta_{I,Brayton\ S-CO_2} = \frac{\dot{W}_{net,Brayton\ S-CO_2}}{\dot{Q}_{RH}} \tag{11}$$

Also, considering the second law of thermodynamics, the exergy efficiency ($\eta_{II,ORC-Brayton\ S-CO_2}$) is calculated, as shown in Equation (12)

$$\eta_{II,ORC-Brayton\ S-CO_2} = \frac{\dot{ED}_{prod}}{\dot{ED}_{fuel}} \tag{12}$$

where $\dot{ED}_{fuel}$ and $\dot{ED}_{prod}$ are the fuel and product exergy rate for the components, which are defined in Table 1.

Table 1. Fuel-Product definition for the ORC and Brayton S-CO$_2$ components.

Cycle	Equipment	Fuel	Product	Loss
ORC	ITC1	$\dot{E}_5$	$\dot{E}_{1AT}-\dot{E}_{3AT}$	-
	P1	$\dot{W}_{P1}$	$\dot{E}_{3AT}-\dot{E}_{2AT}$	-
	ITC2	$\dot{E}_{1AT}-\dot{E}_{2AT}$	$\dot{E}_{1ORC}-\dot{E}_{4ORC}$	-
	T3	$\dot{E}_{1ORC}-\dot{E}_{2ORC}$	$\dot{W}_{T1}$	-
	ITC3	-	-	$\dot{E}_{2A}$
	P2	$\dot{W}_{P2}$	$\dot{E}_{4ORC}-\dot{E}_{3ORC}$	-
Brayton S-CO$_2$	C1	$\dot{W}_{comp}$	$\dot{E}_7 - \dot{E}_6$	-
	T1	$\dot{E}_1 - \dot{E}_2$	$\dot{W}_{t1}$	-
	T2	$\dot{E}_3 - \dot{E}_4$	$\dot{W}_{t2}$	-
	RH	$\dot{E}_8 - \dot{E}_1 + \dot{Q}_s$	$\dot{E}_3 - \dot{E}_2$	-
	HTR	$\dot{E}_4 - \dot{E}_5$	$\dot{E}_8 - \dot{E}_7$	-
	C1	$\dot{W}_{comp}$	$\dot{E}_7 - \dot{E}_6$	-

Thus, the overall thermal efficiency of the integrated system Brayton S-CO$_2$-ORC is a function of net power and heat source, as shown in Equation (13).

$$\eta_{I,overall} = \frac{\dot{W}_{net,Brayton\ S-CO_2} + \dot{W}_{net,ORC}}{\dot{Q}_{RH}} \tag{13}$$

2.4. LCA in the Brayton S-CO$_2$-ORC System

The LCA procedure was adopted to investigate the potential environmental impacts of the components and organic fluids of the *Brayton S-CO$_2$-ORC* in each phase of the life cycle. This procedure is developed according to the ISO 140009 environmental management standards and is supplemented by some steps such as definition, inventory, life cycle impact assessment analysis, results, and interpretation [38].

This analysis was applied in the Barranquilla city (Colombia) in the year 2020. The suggested practical unit is 1 kWh produced by the Brayton S-CO$_2$-ORC system, while the scope of the analysis considers the assembling procedures of materials and divisions of the cycle (construction phase). Also, the operation and maintenance phase was considered as a function of the energy ratio of the equipment. The decommissioning period of the system is likewise considered in the LCA of the thermal system, as shown in Figure 3.

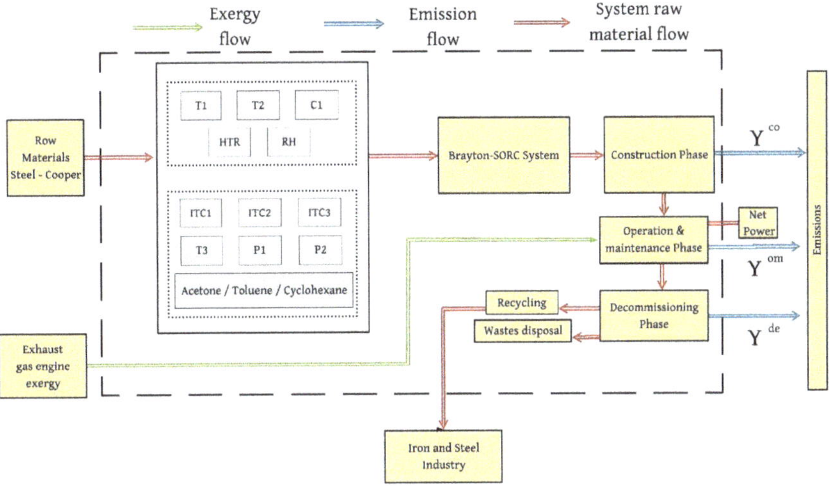

Figure 3. Life cycle assessment system boundary diagram.

Some considerations were adopted to conduct the LCA, such as the Eco-indicator 99 approach, which was utilized in [38]. The environmental impact of the organic working fluid, the components, and the thermal oil are considered in the three phases of the process lifecycle, which are the construction, operation, and maintenance and decommissioning [39]. The normal loss of organic working fluid in an ORC ranges from 0 to 2% of the total filled fluid. So, for a period of 20 years, the working fluid loss considered was 0.5%. As needs are, the liquid loss in the operation phase is 10%, while the organic loss in the decommissioning stage is just 3%. Also, it was assumed that the composition of the thermal oil (Therminol) is 73.5% Diphenyl Oxide for the environmental impact assessment of the thermal oil. Concerning the toxicity of working fluids, in the selection of these, it has been recommended to use nontoxic and nonflammable organic fluids. Therefore, only fluids with classifications A1, B1, A2L, B2L were selected, under ASHRAE standard 34-2001 or the NFPA 704 standard [40,41].

By applying the energy balance in the heat exchangers (ITC1, ITC2, and ITC3), the heat transfer area is obtained with Equation (14)

$$A_i = \frac{\dot{Q}}{\Delta t} \cdot \frac{1}{U} \tag{14}$$

where U is the heat transfer coefficient in kW/m²·K, and Δt is the true temperature difference, determined with the Equation (15)

$$\Delta t = CF_T \cdot LTD \tag{15}$$

where CF_T is the correction factor calculated with the Equation (16), and LTD is the logarithmic mean temperature differences calculated according to Equation (17) [42]

$$CF_T = \frac{\frac{\sqrt{R^2+1}}{R-1} \cdot [\ln(1-S) - \ln(1-RS)]}{\ln \frac{2-S \cdot (R+1-\sqrt{R^2+1})}{2-S \cdot (R+1+\sqrt{R^2+1})}} \tag{16}$$

$$LTD = \frac{\Delta T_{10-1AT} - \Delta T_{11-3AT}}{\ln\left(\frac{\Delta T_{10-1AT}}{\Delta T_{11-3AT}}\right)} \tag{17}$$

where R corresponds to the effectiveness coefficient, and S is the heat power ratio.

The modeling of the printed circuit heat exchangers is carried out according to the mathematical model presented in the literature [31]. This heat exchanger is the Recuperator (HTR) and the Reheater

(RH) in the Brayton cycle, which are fabricated by such technologies as chemical etching and diffusion bonding, where flow channels are imprinted chemically on the metal plates and produce one block by diffusion adhering, as shown in Figure 4.

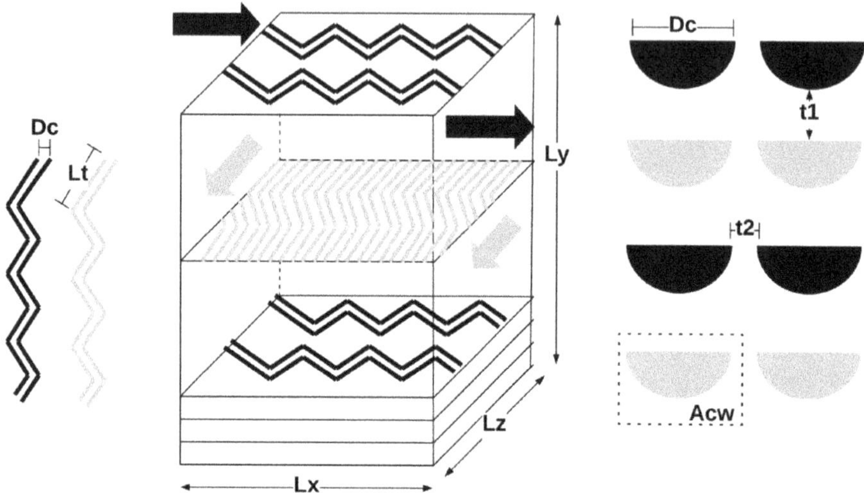

Figure 4. Printed circuit heat exchangers geometrical design.

The heat exchanger is calculated by dividing it into small sub-exchangers (Equation (18)) in which the properties of the fluid, temperature, and pressure are known, and an iterative process is followed until both the heat exchanged and the maximum pressure drop are satisfied. The correlations used for each of the *i* divisions were obtained from the literature [31]

$$N_{Ul} = \begin{cases} 4.089 & \text{if} & Re_i < 2300 \\ 4.089 + \dfrac{Nu_{5000}-4.089}{5000-2300}\cdot\left(Re_i - 2300\right) & \text{if} & 2300 < Re_i < 5000 \\ \dfrac{\left(\frac{f_i}{8}\right)\cdot\left(Re_i-1000\right)\cdot Pr_i}{1+12.7\cdot\left(Pr_i^{\frac{2}{3}}-1\right)\cdot\sqrt{\frac{f_i}{8}}} & \text{if} & Re_i > 5000 \end{cases} \tag{18}$$

where f is the Darcy factor and P_r is the Prandtl number.

The heat transfer coefficient is then calculated using Equation (19)

$$H_i = N_{Ul}\cdot(k/D_{hid}) \tag{19}$$

where k is the conductivity of the exchanger material [W/m·K].

Finally, the overall coefficient U of each element is calculated using Equation (20), and the length of each of the sub-exchangers with Equation (21).

$$U_i = \frac{1}{\frac{1}{H_{h_t}} + \frac{1}{H_{c_t}} + \frac{t}{k}} \tag{20}$$

$$L_i = \frac{Q_i}{P_i\cdot Q_i\cdot\left(T_{h_m} - T_{c_m}\right)} \tag{21}$$

where t is the thickness of the plate, and P_i is the wet parameter.

When calculating the hea°t exchanger, you check that it meets the introduced pressure drop using Equation (22).

$$\Delta P_i = f_i \cdot \left(\frac{L_i}{D_{hid_i}}\right) \cdot \left(\frac{c_i^2}{2}\right) \tag{22}$$

The heat exchanger mass can be calculated using Equation (23)

$$M_i = \rho \cdot A_i \cdot \delta \tag{23}$$

where ρ is the density (steel is 7930 kg/m^3, and copper is 8900 kg/m^3), and δ is the material thickness with a value of 0.002 m [32].

The turbine and pump masses are determined using Equation (24)

$$M_i = \alpha \cdot W_i \tag{24}$$

where W_i is the turbine power produced, or the pump power consumed α is the required quality of the material in kg/kW. For steel, the value for α is 14 kg/kW and 31.22 kg/kW for the pump and the turbine, respectively, while for copper, its value is 35.03 kg/kW and 15.71 kg/kW for the turbine and the pump, respectively [32].

According to Equation (25), the environmental impact of each equipment can be described

$$Y_i = w_i \cdot M_i \tag{25}$$

where w_i is the component coefficient Eco 99.

Therefore, the environmental impact of equipment (Y^{LCA}_k) is obtained using Equation (26) [43]

$$Y^{LCA}_i = Y^{co}_i + Y^{om}_i + Y^{de}_i \tag{26}$$

where Y^{co}_i, Y^{om}_i and Y^{de}_i lead to the environmental impacts of the corresponding phases: construction, operation, and decommissioning.

Finally, the total environmental impacts of components are determined with Equation (27)

$$Y_i = Y^{LCA}_i + Y^{wf}_i \tag{27}$$

where Y^{wf}_i is the effect of the organic fluid volume used in each component, and is a function of the component's exergy destruction, which is determined using Equation (28).

$$Y^{wf}_i = \frac{Y^{wf} \cdot \dot{E}D_{di}}{\dot{E}D_{d-total}} \tag{28}$$

3. Results and Discussion

3.1. System Thermodynamic and Exergy Performance

Figure 5 shows a sensitivity analysis of the thermodynamic and exergy performance indicators of the thermal cycle, based on the inlet temperature of the main turbine (T1) in the Brayton S-CO$_2$-ORC configuration. For the development of the sensitivity analysis, the following considerations were taken into account ORC pressure ratio was 30, evaporator pinch point 25 °C, main turbine inlet temperature (TIT) 750K, Brayton cycle high pressure 25 kPa, Brayton turbine efficiency 93%, ORC turbine 80%, compressor efficiency 89%, pump efficiency 75%, and HTR effectiveness 95%. Figure 5a shows the net power generated with the different organic working fluids simulated in the system, observing that as the temperature increases there is a 36% increase in the net power delivered by the system for a range between 550 °C and 800 °C, showing similar trends among the three organic working fluids studied.

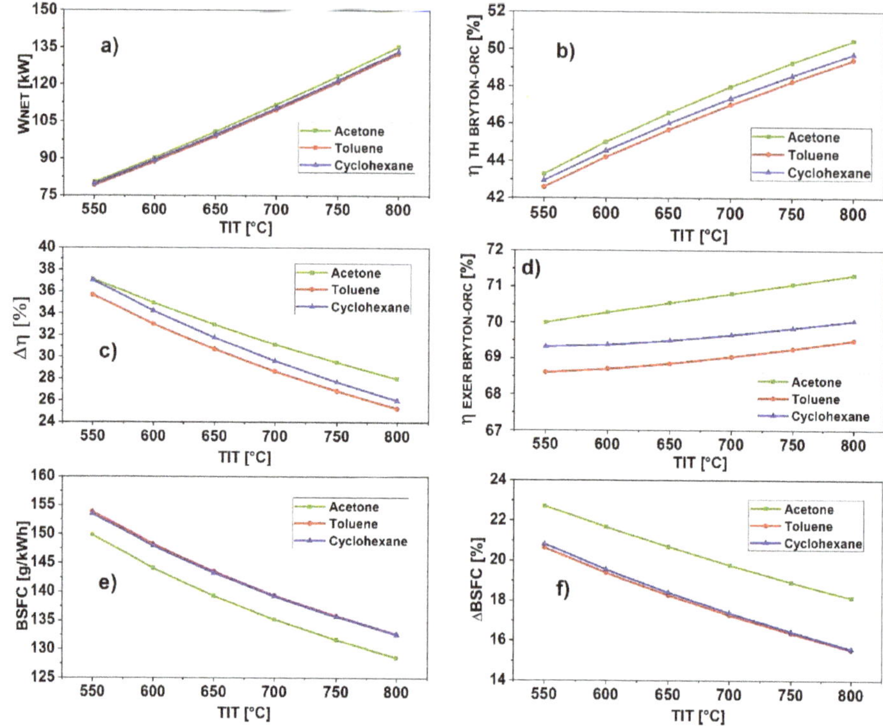

Figure 5. System performance parameters with respect to TIT, (**a**) net power; (**b**) thermal efficiency Brayton-ORC; (**c**) absolute increase in thermal efficiency; (**d**) exergy efficiency Brayton-ORC; (**e**) brake-specific fuel consumption; (**f**) absolute decrease in brake-specific fuel consumption.

In the case of the thermal efficiency of the Brayton S-CO$_2$-ORC integrated system, as shown in Figure 5b, acetone is the fluid with the best thermal performance in the operation of the configurations, this is because its thermo-physical properties benefit the best use of energy and the performance of the system components, reaching a maximum thermal efficiency of 50.44% at 800 °C, this being the best operating condition of this fluid. Therefore, each fluid presents its particular operating conditions that must be studied to optimize the energetic, exergetic, and environmental performance of each proposed alternative [33,34].

Figure 5d shows the total exergetic efficiency of the system studied, where the turbine T1 inlet temperature has a positive influence on all the organic working fluids studied. Toluene and cyclohexane exhibited similar behavior. However, acetone exceeds an exergetic efficiency of 71.4% at a turbine inlet temperature of 800 °C. Through the analysis of this parameter, it contributes to the objective of making more effective use of the nonrenewable energy resource used in the Brayton cycle by establishing the components, modes, and actual amounts of exergy destruction and loss in the integrated system. The values obtained for exergetic efficiency are of great importance in this study since they allow for the design of a more effective integrated Brayton S-CO$_2$-ORC configuration, aimed at the reduction of inefficiencies in these systems. The acetone presented the best average behavior among the fluids in the temperature range studied, concerning toluene and cyclohexane, with a difference of 2.9% and 2.7%, respectively, which is a consequence of the thermal properties of this fluid. However, before implementing these results at the industrial level, advanced economic analyses and exertion of these systems must be carried out [35], in addition to thermo-economic studies to determine the technical and economic feasibility of this solution [36].

The high turbine pressure in the system is another relevant operational parameter that impacts on the energy and exergy efficiency of the integrated system; however, this variable has less impact than the main turbine inlet temperature (T1), producing less variability in the performance parameters, as can be seen in Figure 6. For the performance of the net power, it presents less efficiency in its behavior concerning the performance affected by the inlet temperature (Figure 5a), presenting an approximate increase of 36% beside 9% of the energy generated by the system.

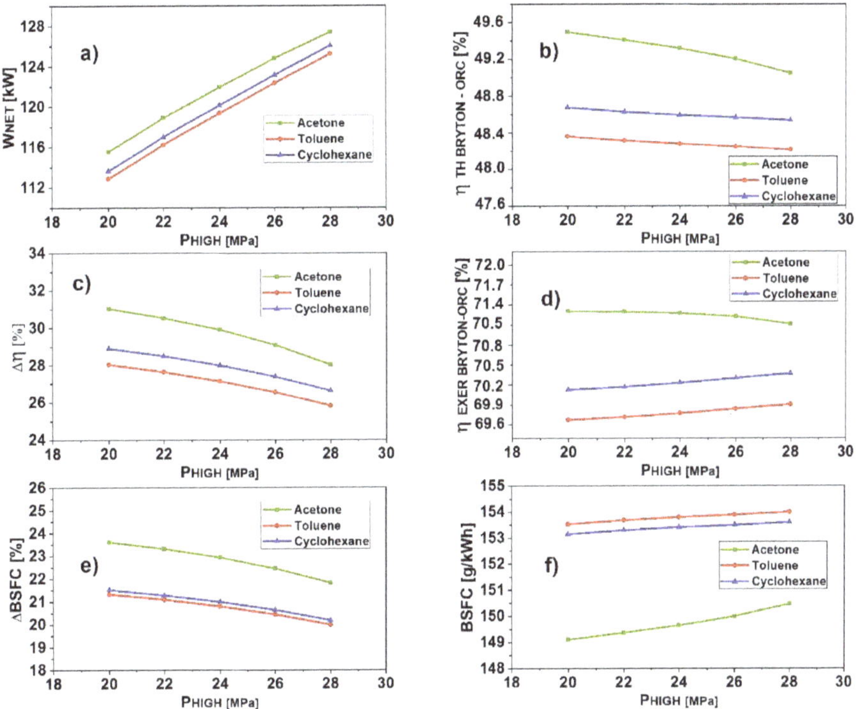

Figure 6. System performance parameters with respect to P_{HIGH}, (**a**) net power; (**b**) thermal efficiency Brayton-ORC; (**c**) absolute increase in thermal efficiency; (**d**) exergy efficiency Brayton-ORC; (**e**) brake-specific fuel consumption; (**f**) absolute decrease in brake-specific fuel consumption.

On the other hand, Figure 6e shows the decrease in the specific fuel consumption for the three fluids studied in the integrated configuration of the Brayton S-CO$_2$-ORC system, with acetone showing the best performance of the three fluids due to its better specific consumption reduction. This result is closed near the other parameter calculated. This result implies a significant reduction in the system operating costs. It allows better use of resources, improving its performance between the energy input and the power produced at high temperatures.

The exergetic efficiency behavior for each component of the system under the three organic working fluids is shown in Figure 7. From the results. The lower exergy efficiency is presented in the thermal oil pump (P1), with a 10.98% (Toluene) and 11.06% (Cyclohexane), while the efficiency in the organic fluid pump (P2) was 77.6% for the three fluids. These results are because of the higher-pressure ratios required to pump the thermal oil, which implies higher irreversibilities for heat transfer in this component. Thus, a thermo-hydraulic design should be proposed for both the evaporator (ITC2) and the shell and tube heat exchanger (ITC1) with the lowest pressure drop, and highest heat transfer.

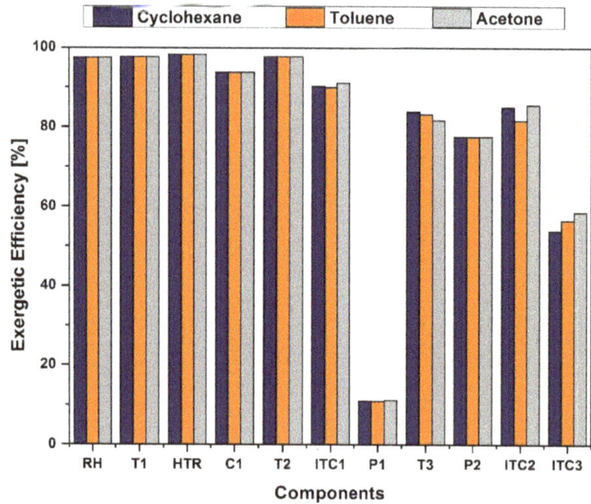

Figure 7. Exergetic efficiency of the Brayton S-CO_2-ORC components.

For the operating conditions studied, the turbine 1 (T1) and the turbine (T2) are the components that present the best behavior, presenting exergetic efficiencies of 98% in the three fluids used to analyze this configuration. Therefore, they present minor irreversibilities and allow better use of the energy in the system.

3.2. Exergy Destruction

To develop an exergy destruction analysis of the system, it was necessary to apply the exergy balance for each of the components as shown in Figure 8, where the exergy destruction fraction of each component is evaluated at different inlet turbine temperature ranging from 550 °C to 800 °C. For this analysis, the operational considerations in Section 3.1 have been taken to determine exergy destruction. The component with the minimum exergy destroyed is the pump (P2) compared to the heat exchanger components that have the greatest exergy destruction in the system. Thus, the ITC2, which is the component with the greatest exergy destroyed with values ranging from 3.93 kW to 9.75 kW for the temperature range evaluated, showing a decrease between the value of exergy destroyed from the base condition of 11% and an increase in exergy destroyed at the temperature of 800 °C of 3%. Although technological improvements do not translate into significant improvements in exergetic efficiency, this result can be improved if ITC1 is designed with rational energy use and sustainability criteria, that leads to sensible improvements in how heat transfer is performed in this type of exchanger, with the aim to obtain more compact, economical and efficient equipment.

The thermal oil pump (P1), and the organic fluid pump (P2) were the components that presented minimum exergy destruction. An alternative to having equipment with less exergy destruction would be to propose pumping systems with higher exergy efficiency, which would imply an increase in the net power produced by the cycle because the pump will consume less energy. However, the pump power is less than that produced by the turbine, which is the reason for the lower contribution of the exergy destroyed and the isentropic efficiency on the overall thermal cycle performance, as the contribution of these is almost indistinguishable when compared with the exergy destroyed from the other components of the cycle at the different inlet turbine temperatures.

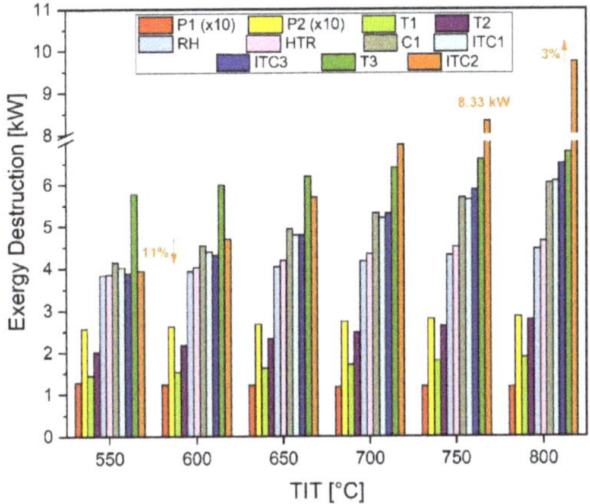

Figure 8. Components destroyed exergy with respect to turbine inlet temperature (TIT).

The influence of the P_{HIGH} on the destroyed exergy by components is presented in Figure 9, where the ITC2 is the component with the highest irreversibility in the configuration studied. This trend of the exergy destruction continues as the high-pressure turbine decreases, obtaining approximately 8.67 kW of exergy destroyed in the system by this component when the pressure is 20 MPa, which is reflected in a significant energy loss in this component due to the large size of this heat exchanger. Therefore, the operating conditions in the Brayton cycle at the input of operation, although it makes the cycle deliver more power, at the time of coupling with ORC cycle, show important irreversibilities by heat transfer in the thermal circuit of coupling due to operational limitations to ensure thermal stability on the thermal oil and organic fluid. These results can be compared with the results made by the authors in the following references, where they use other working fluids and different models and heat recovery systems [44,45].

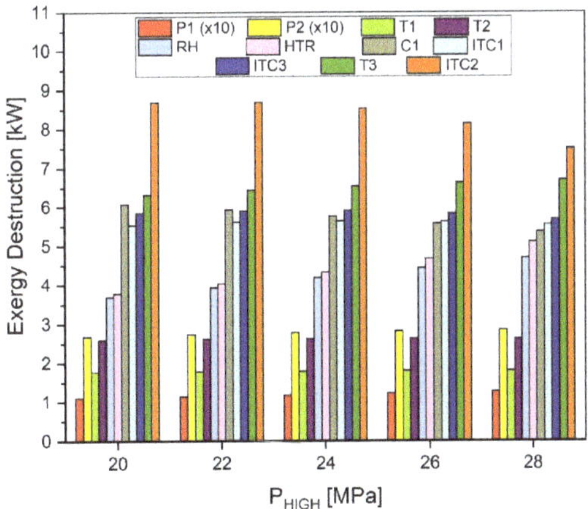

Figure 9. Components destroyed exergy with respect to P_{HIGH}.

Similarly, the exergy destruction in the HTR used in the system increases as the P_{HIGH} increases from 20 MPa to 28 MPa, being this the second component with more impact in the process, reaching a value of 5.10 kW in the exergy destruction at the maximum operating condition. These results are due to the presence of greater irreversibilities in the thermal source, as it is required that the organic working fluid reaches a higher pressure and temperature; therefore, this variable must be considered as an objective variable in energy and exergetic optimization to obtain competitive efficiencies for the system operating with the energy source under study.

3.3. Life Cycle Assessment

Based on the thermodynamic parameters of each state of the Brayton S-CO$_2$-ORC integrated system, the energy and exergetic parameters of the organic components and organic fluids are shown in Table 2. In the proposed system, a lifetime of 20 years was considered [46], in which the components and working fluids will have 7446 working hours [47,48].

Table 2. Exergy efficiency, destroyed exergy, destroyed exergy ratio of each equipment, and heat exchanger area.

Parameter	ITC 1			ITC2		
	Cyclohexane	Toluene	Acetone	Cyclohexane	Toluene	Acetone
$\dot{Q}_i$ [kW]	182.00	166.62	181.84	165.08	165.08	165.08
A_i [m^2]	88.70	88.70	88.70	18.73	16.86	17.51
ε_i [%]	90.27	89.96	91.08	85.02	81.71	85.54
$\dot{ED}_k$ [kW]	6.15	5.17	5.64	8.55	10.41	8.33
$y_{D,k}$ [%]	13.15	11.09	12.32	18.28	22.32	18.20

Parameter	ITC3		
	Cyclohexane	Toluene	Acetone
$\dot{Q}_i$ [kW]	138.56	139.36	136.86
A_i [m^2]	10.06	8.58	6.46
ε_i [%]	53.85	56.53	58.52
$\dot{ED}_k$ [kW]	7.69	6.74	5.88
$y_{D,k}$ [%]	16.44	14.44	12.86

Parameter	RHR			HTR		
	Cyclohexane	Toluene	Acetone	Cyclohexane	Toluene	Acetone
$\dot{Q}_i$ [kW]	250.46	250.46	250.46	446.79	446.79	446.79
A_i [m^2]	176.24	176.24	176.24	15.66	15.66	15.66
ε_i [%]	97.57	97.57	97.57	98.30	98.30	98.30
$\dot{ED}_k$ [kW]	4.31	4.31	4.31	4.50	4.50	4.50
$y_{D,k}$ [%]	9.21	9.25	9.42	9.61	9.65	9.83

Parameter	T1			T2		
	Cyclohexane	Toluene	Acetone	Cyclohexane	Toluene	Acetone
$\dot{W}_i$ [kW]	76.82	76.82	76.82	108.86	108.86	108.86
ε_i [%]	97.71	97.71	97.71	97.64	97.64	97.64
$\dot{ED}_k$ [kW]	1.80	1.80	1.80	2.63	2.63	2.63
$y_{D,k}$ [%]	3.84	3.86	3.93	5.61	5.63	5.74

Parameter	T3			C1		
	Cyclohexane	Toluene	Acetone	Cyclohexane	Toluene	Acetone
$\dot{W}_i$ [kW]	27.15	25.94	29.50	90.39	90.39	90.39
ε_i [%]	83.88	83.22	81.74	93.72	93.72	93.72
$\dot{ED}_k$ [kW]	5.22	5.23	6.59	5.68	5.68	5.68
$y_{D,k}$ [%]	11.14	11.21	14.40	12.14	12.18	12.41

Parameter	P1			P2		
	Cyclohexane	Toluene	Acetone	Cyclohexane	Toluene	Acetone
$\dot{W}_i$ [kW]	0.14	0.14	0.14	0.62	0.21	1.27
ε_i [%]	11.06	10.98	11.25	77.62	77.60	77.66
$\dot{ED}_k$ [kW]	0.13	0.13	0.12	0.14	0.05	0.28
$y_{D,k}$ [%]	0.27	0.27	0.27	0.30	0.10	0.62

Parameter	Brayton S-CO2-ORC		
	Cyclohexane	Toluene	Acetone
$\dot{W}_{net}$ [kW]	121.67	120.87	123.38
$\eta_{I,Brayton\ S-CO_2-ORC}$ [%]	48.58	48.26	49.26
$\Delta\eta$ [%]	27.70	26.86	29.49
$\eta_{II,Brayton\ S-CO_2-ORC}$ [%]	70.27	69.81	71.25

The results allow for identifying that the RHR is one of the components that presents greater environmental impacts due to their heat transfer area in comparison with the rest of the components of the system. Also, the RHR presents the greatest exergy destroyed, with values of 46% (cyclohexane), 32% (toluene), and 32% (acetone).

The masses in the different phases of the life cycle of the Brayton S-CO$_2$-ORC system are presented in Table 3. The LCA methodology was developed to obtain the environmental impacts of each component of the system under the three phases of the lifetime, which are the construction, operation and decommissioning, as shown in Table 4, selecting steel as the material. Table A1 shows the results when copper is proposed as the construction material for the devices.

Table 3. Organic fluids and thermal oil masses in each life cycle phase.

Fluid	LCA Phase		
	Construction	Operation	Decommissioning
Cyclohexane	151.2078	15.1208	4.0826
Toluene	144.4687	14.4469	3.9007
Acetone	164.3074	16.4307	4.4363
Thermal Oil	184.4000	18.44	4.9788

Table 4. Results of Life Cycle Assessment for each component selecting steel as the material.

Organic Fluid	Equipment	Material	w [mPts/kg]	Quality [kg]	Y^{co} [mPts]	Y^{om} [mPts]	Y^{de} [mPts]	Y [mPts]
	ITC1	Steel	86	1407	127,037	0	6049	133,086
	ITC2	Steel	86	297	26,823	0	1277	28,100
	ITC3	Steel	86	159	14,402	0	686	15,088
	RHR	Steel	86	2795	252,404	0	12,019	264,423
	HTR	Steel	86	248	22,428	0	1068	23,496
	T1	Steel	86	2398	216,579	0	10,313	226,893
Cyclohexane	T2	Steel	86	3399	306,903	0	14,614	321,518
	T3	Steel	86	848	76,531	0	3644	80,176
	C1	Steel	86	2822	254,838	0	12,135	266,973
	P1	Steel	86	2	178	0	8	186
	P2	Steel	86	9	789	0	38	826
	Thermal Oil	Therminol	46,467	184	856,8515	856,851	231,350	9,656,716
	Fluid	Cyclohexane	2639	151	399,022	39,902	10,774	449,698
	ITC1	Steel	86	1407	127,037	0	6049	133,086
	ITC2	Steel	86	267	24,139	1	1149	25,290
	ITC3	Steel	86	136	12,291	2	585	12,879
	RHR	Steel	86	2795	252,404	3	12,019	264,426
	HTR	Steel	86	248	22,428	4	1068	23,500
	T1	Steel	86	2398	216,579	5	10,313	226,898
Toluene	T2	Steel	86	3399	306,903	6	14,614	321,524
	T3	Steel	86	810	73,121	7	3482	76,609
	C1	Steel	86	2822	254,838	8	12,135	266,981
	P1	Steel	86	2	178	9	8	195
	P2	Steel	86	3	270	10	13	293
	Thermal Oil	Therminol	46,467	184	8,568,515	856,851	231,350	9,656,716
	Fluid	Toluene	2680	144	387,147	10,941	10,453	408,541
	ITC1	Steel	86	1407	127,037	0	6049	133,086
	ITC2	Steel	86	278	25,081	1	1194	26,277
	ITC3	Steel	86	103	9257	2	441	9700
	RHR	Steel	86	2795	252,404	3	12,019	264,426
	HTR	Steel	86	248	22,428	4	1068	23,500
	T1	Steel	86	2398	216,579	5	10,313	226,898
Acetone	T2	Steel	86	3399	306,903	6	14,614	321,524
	T3	Steel	86	921	83,162	7	3960	87,129
	C1	Steel	86	2822	254,838	8	12,135	266,981
	P1	Steel	86	2	178	9	8	195
	P2	Steel	86	18	1612	10	77	1699
	Thermal Oil	Therminol	46,467	184	8,568,515	856,851	231,350	9,656,716
	Fluid	Acetone	5426	164	891,450	89,145	24,069	1,004,664

For the three organic working fluids (cyclohexane, toluene, and acetone), the results of the assessment of the environmental impact of the system are presented in Figure 10, where the material of the components is the steel and copper. The components with the greatest environmental impacts are the main turbine (T1) and the secondary turbine T2 of the Brayton S-CO$_2$ cycle, with values of 226,893 mPts and 321,518 mPts, respectively, when steel is selected as the material. With this, the T1 and T2 turbines obtain, respectively, percentage values of 18.19% and 25.78% (cyclohexane), 20.54% and 29.10% (toluene), 20.36% and 28.85% (acetone). In the ORC, the ITC1 is the component with the greatest environmental impact, with percentage values of 10.67% (cyclohexane), 12.05% (toluene), and 11.94% (acetone) concerning the environmental impacts of all components. When the material is copper, there is a decrease in the percentage values of the T1 and T2 turbines due to the increase in environmental impacts in the ITC1, which obtains values of 11.45% (cyclohexane), 12.85% (toluene) and 12.94% (acetone).

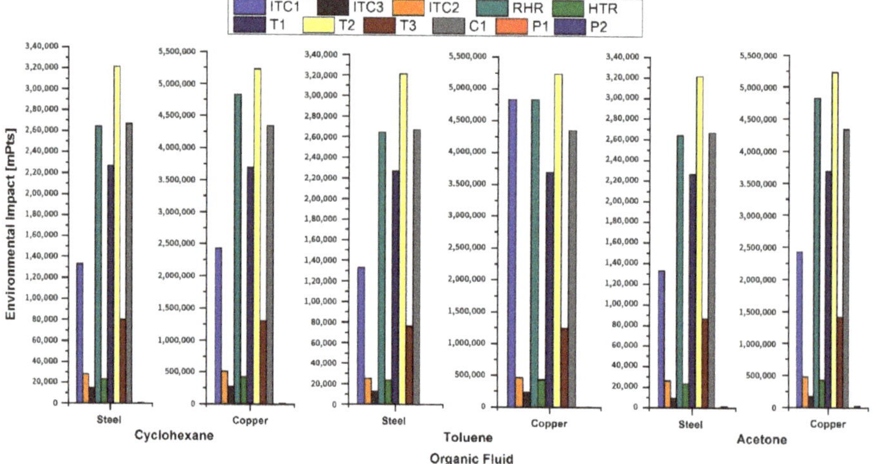

Figure 10. Environmental impact of each equipment with respect to the organic fluids and materials.

In general, the environmental impacts of each equipment when steel is selected as material steel are lower than those of the components with copper because in the methodology applied the Eco99 coefficient is higher than that of steel. Therefore, considering sustainability criteria complementary to the energy and exergetic aspects studied, the organic working fluid and material in which the greatest exergetic opportunities for improvement are found can be selected to obtain the least environmental impact.

Regarding the environmental impacts of working fluids and thermal oil, the results are presented in Figure 11 where the thermal oil has a very large environmental impact on the organic fluids, representing 83% of the environmental impacts compared to cyclohexane (4%), toluene (4%), and acetone (9%), respectively. Among the organic fluids, because it has a higher Eco99 coefficient, acetone has greater environmental impacts than cyclohexane and toluene, and the suggested material is the steel, which is a result close to that reported in the literature for the case of the ORC as heat recovery [37]. Therefore, with the ORC being built with this material and operating with this fluid, it can be widely used for heat recovery from natural gas generation engines [38].

The results show that the greatest differences in LCA are obtained in the environmental impacts in the construction phase, where acetone presents better results for this system, which coincides with the results obtained from the analysis of the energy and exergetic indicators.

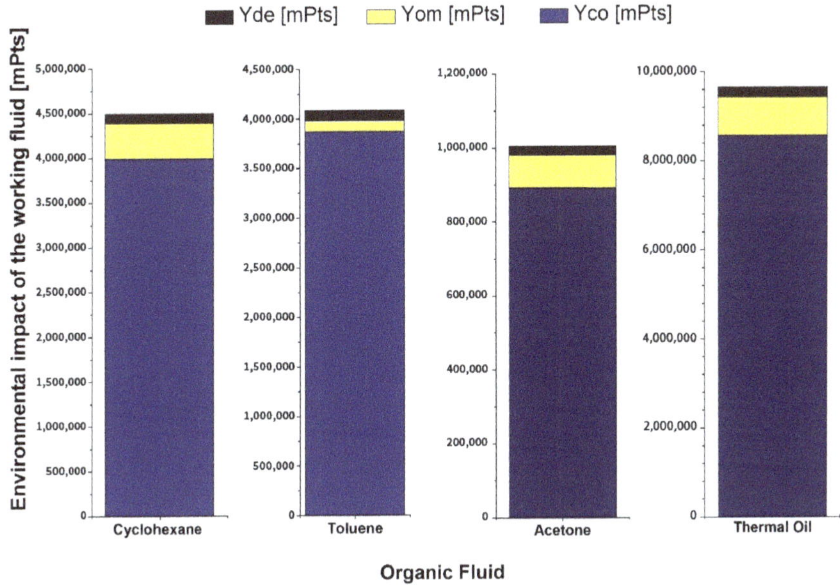

Figure 11. Environmental impact of the working fluid.

4. Conclusions

The main contribution of this study was to analyze the energy, exergy, and environmental performance of an integrated Brayton S-CO$_2$-ORC, where some performance indicators and the potential environmental impact were studied using Toluene, Acetone, and Cyclohexane as the organic working fluids. Therefore, a complementary assessment was developed based on the energy, exergetic, and environmental analysis to determine the best behavior of the components in this cycle, and the effect of the relevant operating conditions.

For identifying the best performance of the components that integrate this system, the study evaluates exergetic and energy parameters such as the exergetic and overall thermal efficiency under the different organic working fluids used in this study. In these parameters, it is possible to evaluate the power generated by the integrated system for each fluid operating in the Brayton S-CO$_2$-ORC configuration, where the acetone presented the best behavior with respect to toluene and cyclohexane. The performance of the system for the operating conditions studied found that turbine 1 (T1) and the turbine (T2) are the components that present the best exergetic efficiencies with 97%. Therefore, the acetone offers minor irreversibilities and allows for better performance in the ORC as a bottoming cycle from the Brayton cycle.

Conversely, another relevant parameter studied is the exergy destroyed of each component, presenting the pump (P2) least performance compared to the heat exchanger components that have the most considerable exergy destruction in the system, thus observing that ITC1 is the component with the most exergy destroyed, with values ranging from 75 kW to 127 kW for the temperature range evaluated. So, this component presents greater opportunities for energy improvement, and therefore environmental improvement, since its potential environmental impacts are based on the heat transfer area of this equipment.

For the three working fluids (cyclohexane, toluene and acetone), the results of the potential environmental impacts of the system were studied using the steel and copper as construction materials for the components. The components with the greatest environmental impacts are the main turbine (T1), and the secondary turbine (T2) of the Brayton S-CO$_2$ cycle, with values of 226 Pts and 321

Pts, respectively. The T1 and T2 turbines obtained a percentage value of 18.19% and 25.78% of the total environmental impact with cyclohexane, 20.54%, and 29.10% with toluene, 20.36% and 28.85% with acetone. Also, the ITC1 in the ORC cycle is the component with the greatest environmental impact, with a percentage value of 10.67% (cyclohexane), 12.05% (toluene) and 11.94% (acetone).

In general, the environmental impacts of the components with steel are lower than those of the components with copper, because the methodology applied suggests a higher Eco99 coefficient than steel. Therefore, through this methodology, the organic working fluid and the material in which the most significant opportunities for improvement are found can be selected to obtain the smallest environmental impact.

Among the organic fluids studied, acetone has lower potential environmental impacts than cyclohexane and toluene, which is a consequence of the Eco99 coefficient. However, some safety and health consideration such be considered to implement the ORC alternative industrially as the bottoming system from the Brayton cycle.

Finally, this study allows for evaluating the performance of the combined cycles for applying this technology in industries for generating energy and net power. Acetone is the fluid with the best thermodynamic and environmental performance results on this configuration because of their thermal properties, giving an option for other studies to proposed an eco-design of this and obtain better exergy and environmental results according to other performance parameters.

Author Contributions: Conceptualization: E.E.B.; Methodology: G.V.O. and J.D.F.; Software: E.E.B., G.V.O., and J.D.F.; Validation: E.E.B., and J.D.F.; Formal Analysis: E.E.B., G.V.O., and J.D.F.; Investigation: E.E.B., G.V.O., and J.D.F.; Resources: G.V.O. and J.D.F.; Writing—Original Draft Preparation: G.V.O.; Writing—Review and Editing: G.V.O. and J.D.F.; Funding Acquisition: G.V.O., and J.D.F. All authors have read and agreed to the published version of the manuscript.

Funding: This work was supported by the Universidad del Atlántico, and Universidad Francisco de Paula Santander in Ocaña - Norte de Santander.

Acknowledgments: This research was supported by the Mechanical Engineering Program of Universidad del Atlántico. The Kai Research Group supports G. Valencia and J. Duarte.

Conflicts of Interest: The authors declare no conflict of interest.

Nomenclature

LCA	Life Cycle Assessment
ORC	Organic Rankine Cycle
$\dot{Q}$	Heat rate [kW]
$\dot{W}$	Power [kW]
$\dot{m}$	Mass flow rate [kW]
h	Enthalpy [kJ/kg·K]
s	Specific entropy [kJ/kg·K]
ex	Exergy rate
$\dot{E}_D$	Exergy destruction rate
η	Efficiency
M	Mass [kg]
A	Area [m]
δ	Thickness [m]
CF_T	Correction Factor
Y_i	Environmental Impact [mPts]
Co	Construction
Om	Operation
De	decommissioning
Wf	Working fluid
PPE	Pinch Point of Evaporator

Appendix A

The LCA results in each component of the proposed configurations using copper as the material are presented in Table A1.

Table A1. Results of Life Cycle Assessment for each component selecting copper as the material.

Organic Fluid	Equipment	Material	w [mPts/kg]	Quality [kg]	Y^{co} [mPts]	Y^{om} [mPts]	Y^{de} [mPts]	Y [mPts]
Cyclohexane	ITC1	Copper	1400	1579	2,321,003	0	110,524	2,431,527
	ITC2	Copper	1400	333	490,057	0	23,336	513,393
	ITC3	Copper	1400	179	263,130	0	12,530	275,660
	RHR	Copper	1400	3137	4,611,496	0	219,595	4,831,091
	HTR	Copper	1400	279	409,760	0	19,512	429,272
	T1	Copper	1400	2398	3525712	0	167,891	3,693,603
	T2	Copper	1400	3399	4,996,099	0	237,909	5,234,008
	T3	Copper	1400	848	1,245,860	1	59,327	1,305,187
	C1	Copper	1400	2822	4,148,525	2	197,549	434,6076
	P1	Copper	1400	2	2894	3	138	3035
	P2	Copper	1400	9	12,842	4	612	13,457
	Thermal Oil	Therminol	46,467	184	8,568,515	856,851	231,350	9,656,716
	Fluid	Cyclohexane	2639	149	392,031.28	39,203.13	10,585	441,819
Toluene	ITC1	Copper	1400	3137	4,611,495.84	0	219,595	4,831,091
	ITC2	Copper	1400	300	441,035.62	1	21,002	462,038
	ITC3	Copper	1400	153	224,568.70	2	10,694	235,264
	RHR	Copper	1400	3137	4,611,495.84	3	219,595	4,831,094
	HTR	Copper	1400	279	409,759.56	4	19,512	429,276
	T1	Copper	1400	2398	3,525,711.97	5	167,891	3,693,608
	T2	Copper	1400	3399	4,996,098.50	6	237,909	5,234,014
	T3	Copper	1400	810	1,190,334.01	7	56,683	1,247,024
	C1	Copper	1400	2822	4,148,525.01	8	197,549	4,346,082
	P1	Copper	1400	2	2895.92	9	138	3043
	P2	Copper	1400	3	4398.53	10	209	4618
	Thermal Oil	Therminol	46,467	184	8,568,515	856,851	231,350	9,656,716
	Fluid	Toluene	2680	144	386,282.14	10,748.90	10,430	407,461
Acetone	ITC1	Copper	1400	1579	2,321,002.70	0	110,524	2,431,527
	ITC2	Copper	1400	312	458,241.95	1	21,821	480,064
	ITC3	Copper	1400	115	169,131.42	2	8054	177,187
	RHR	Copper	1400	3137	4,611,495.84	3	219,595	4831094
	HTR	Copper	1400	279	409,759.56	4	19,512	429,276
	T1	Copper	1400	2398	3,525,711.97	5	167,891	3,693,608
	T2	Copper	1400	3399	4,996,098.50	6	237,909	5,234,014
	T3	Copper	1400	921	1,353,793.03	7	64,466	1,418,266
	C1	Copper	1400	2822	4,148,525.01	8	197,549	4,346,082
	P1	Copper	1400	2	2889.66	9	138	3036
	P2	Copper	1400	18	26,239.27	10	1249	27,499
	Thermal Oil	Therminol	46,467	184	8,568,515	856,851	231,350	9,656,716
	Fluid	Acetone	5426	170	924,019.85	92,401.98	24,949	1,041,370

References

1. Diaz, G.A.; Forero, J.D.; Garcia, J.; Rincon, A.; Fontalvo, A.; Bula, A.; Padilla, R.V. Maximum power from fluid flow by applying the first and second laws of thermodynamics. *J. Energy Resour. Technol.* **2017**, *139*, 032903. [CrossRef]
2. Ramírez, R.; Gutiérrez, A.S.; Eras, J.J.C.; Valencia, K.; Hernández, B.; Forero, J.D. Evaluation of the energy recovery potential of thermoelectric generators in diesel engines. *J. Clean. Prod.* **2019**, *241*, 118412. [CrossRef]
3. Angelino, G. Carbon Dioxide Condensation Cycles. *J. Eneg. Power* **1968**, 287–295. [CrossRef]
4. Dostal, V.; Driscoll, M.J.; Hejzlar, P.A. Supercritical Carbon Dioxide Cycle for Next Generation Nuclear Reactors. Ph.D. Thesis, Massachusetts Institute of Technology, Cambridge, UK, March 2004.
5. Abrosimov, K.A.; Baccioli, A.; Bischi, A. Techno-economic analysis of combined inverted Brayton—Organic Rankine cycle for high-temperature waste heat recovery. *Energy Convers. Manag. X* **2019**, *3*, 100014. [CrossRef]

6. Guo, Z.; Zhao, Y.; Zhu, Y.; Niu, F.; Lu, D. Optimal design of supercritical CO_2 power cycle for next generation nuclear power conversion systems. *Prog. Nucl. Energy* **2018**, *108*, 111–121. [CrossRef]

7. Padilla, R.V.; Soo Too, Y.C.; Benito, R.; Stein, W. Exergetic analysis of supercritical CO_2 Brayton cycles integrated with solar central receivers. *Appl. Energy* **2015**, *148*, 348–365. [CrossRef]

8. Padilla, R.V.; Benito, R.G.; Stein, W. An Exergy Analysis of Recompression Supercritical CO_2 Cycles with and without Reheating. *Energy Procedia* **2015**, *69*, 1181–1191. [CrossRef]

9. Glatzmaier, G.C.; Turchi, C.S. Supercritical CO_2 as a Heat Transfer and Power Cycle Fluid for CSP Systems. In Proceedings of the ASME 2009 3rd International Conference on Energy Sustainability collocated with the Heat Transfer and InterPACK09 Conferences, San Francisco, CA, USA, 19–23 July 2009; pp. 673–676.

10. Hinze, J.F.; Nellis, G.F.; Anderson, M.H. Cost comparison of printed circuit heat exchanger to low cost periodic flow regenerator for use as recuperator in a s-CO_2 Brayton cycle. *Appl. Energy* **2017**, *208*, 1150–1161. [CrossRef]

11. Sharan, P.; Neises, T.; Turchi, C. Thermal desalination via supercritical CO_2 Brayton cycle: Optimal system design and techno-economic analysis without reduction in cycle efficiency. *Appl. Ther. Eng.* **2019**, *152*, 499–514. [CrossRef]

12. Park, J.H.; Park, H.S.; Kwon, J.G.; Kim, T.H.; Kim, M.H. Optimization and thermodynamic analysis of supercritical CO_2 Brayton recompression cycle for various small modular reactors. *Energy* **2018**, *160*, 520–535. [CrossRef]

13. Li, H.; Zhang, Y.; Zhang, L.; Yao, M.; Kruizenga, A.; Anderson, M. PDF-based modeling on the turbulent convection heat transfer of supercritical CO_2 in the printed circuit heat exchangers for the supercritical CO_2 Brayton cycle. *Int. J. Heat Mass Transf.* **2016**, *98*, 204–218. [CrossRef]

14. Ahn, Y.; Bae, S.J.; Kim, M.; Cho, S.K.; Baik, S.; Lee, J.I.; Cha, J.E. Review of supercritical CO 2 power cycle technology and current status of research and development. *Nucl. Eng. Technol.* **2015**, *47*, 647–661. [CrossRef]

15. Musgrove, G.; Sullivan, S.; Shiferaw, D.; Fourspring, P. Heat exchangers. In *Fundamentals and Applications of Supercritical Carbon Dioxide (SCO_2) Based Power Cycles*; Elsevier Ltd.: Amsterdam, The Netherlands, 2017; pp. 217–244.

16. Jiang, Y.; Liese, E.; Zitney, S.E.; Bhattacharyya, D. Optimal design of microtube recuperators for an indirect supercritical carbon dioxide recompression closed Brayton cycle. *Appl. Energy* **2018**, *216*, 634–648. [CrossRef]

17. Jiang, Y.; Liese, E.; Zitney, S.E.; Bhattacharyya, D. Design and dynamic modeling of printed circuit heat exchangers for supercritical carbon dioxide Brayton power cycles. *Appl. Energy* **2018**, *231*, 1019–1032. [CrossRef]

18. Chen, J.; Liu, Y.; Lu, X.; Ji, X.; Wang, C. Designing heat exchanger for enhancing heat transfer of slurries in biogas plants. *Energy Procedia* **2019**, *158*, 1288–1293. [CrossRef]

19. Pidaparti, S.R.; Anderson, M.H.; Ranjan, D. Experimental Investigation of thermal-hydraulic performance of discontinuous fin printed circuit heat exchangers for Supercritical CO_2 power cycles. *Exp. Ther. Fluid Sci.* **2019**, *106*, 119–129. [CrossRef]

20. Colonna, P.; van Putten, H. Dynamic modeling of steam power cycles. Part I-Modeling paradigm and validation. *Appl. Ther. Eng.* **2007**, *27*, 467–480. [CrossRef]

21. Van Putten, H.; Colonna, P. Dynamic modeling of steam power cycles: Part II—Simulation of a small simple Rankine cycle system. *Appl. Ther. Eng.* **2007**, *27*, 2566–2582. [CrossRef]

22. Chien, N.B.; Jong-Taek, O.; Asano, H.; Tomiyama, Y. Investigation of experiment and simulation of a plate heat exchanger. *Energy Procedia* **2019**, *158*, 5635–5640. [CrossRef]

23. Liu, Y.; Wang, Y.; Huang, D. Supercritical CO_2 Brayton cycle: A state-of-the-art review. *Energy* **2019**, *189*, 115900. [CrossRef]

24. Mohammadkhani, F.; Shokati, N.; Mahmoudi, S.M.S.; Yari, M.; Rosen, M.A. Exergoeconomic assessment and parametric study of a Gas Turbine-Modular Helium Reactor combined with two Organic Rankine Cycles. *Energy* **2014**, *65*, 533–543. [CrossRef]

25. Turchi, C.S.; Ma, Z.; Neises, T.; Wagner, M. Thermodynamic Study of Advanced Supercritical Carbon Dioxide Power Cycles for High Performance Concentrating Solar Power Systems. In Proceedings of the ASME 2012 6th International Conference on Energy Sustainability Collocated with the ASME 2012 10th International Conference on Fuel Cell Science, Engineering and Technology, San Diego, CA, USA, 23–26 July 2012; pp. 375–383.

26. Le Moullec, Y. Conceptual study of a high efficiency coal-fired power plant with CO_2 capture using a supercritical CO_2 Brayton cycle. *Energy* **2013**, *49*, 32–46. [CrossRef]

27. Olumayegun, O.; Wang, M.; Oko, E. Thermodynamic performance evaluation of supercritical CO_2 closed Brayton cycles for coal-fired power generation with solvent-based CO_2 capture. *Energy* **2019**, *166*, 1074–1088. [CrossRef]

28. Liang, Y.; Bian, X.; Qian, W.; Pan, M.; Ban, Z.; Yu, Z. Theoretical analysis of a regenerative supercritical carbon dioxide Brayton cycle/organic Rankine cycle dual loop for waste heat recovery of a diesel/natural gas dual-fuel engine. *Energy Convers. Manag.* **2019**, *197*, 111845. [CrossRef]

29. Kao, S.; Gibbs, J.; Hejzlar, P. Dynamic Simulation and Control of a Supercritical CO_2 Power Conversion System for Small Light Water Reactor Applications. In Proceedings of the Supercritical CO_2 Power Cycle Symposium, Troy, NY, USA, 29–30 April 2009.

30. Uusitalo, A.; Ameli, A.; Turunen-Saaresti, T. Thermodynamic and turbomachinery design analysis of supercritical Brayton cycles for exhaust gas heat recovery. *Energy* **2019**, *167*, 60–79. [CrossRef]

31. Danieli, P.; Rech, S.; Lazzaretto, A. Supercritical CO2 and air Brayton-Joule versus ORC systems for heat recovery from glass furnaces: Performance and economic evaluation. *Energy* **2019**, *168*, 295–309. [CrossRef]

32. Thakar, R.; Bhosle, S.; Lahane, S. Design of Heat Exchanger for Waste Heat Recovery from Exhaust Gas of Diesel Engine. *Procedia Manuf.* **2018**, *20*, 372–376. [CrossRef]

33. Valencia, G.; Isaza-Roldan, C.; Forero, J. Economic and Exergo-Advance Analysis of a Waste Heat Recovery System based on Regenerative Organic Rankine Cycle under Organic Fluids with Low Global Warming Potential. *Energies* **2020**, *13*, 1317.

34. Ochoa, G.V.; Peñaloza, C.A.; Forero, J.D. Thermo-economic assessment of a gas microturbine-absorption chiller trigeneration system under different compressor inlet air temperatures. *Energies* **2019**, *12*, 4643. [CrossRef]

35. Valencia, G.; Peñaloza, C.; Rojas, J. Thermoeconomic Modelling and Parametric Study of a Simple ORC for the Recovery of Waste Heat in a 2 MW Gas Engine under Different Working Fluids. *Appl. Sci.* **2019**, *9*, 4017. [CrossRef]

36. Ochoa, G.V.; Isaza-Roldan, C.; Forero, J.D. A phenomenological base semi-physical thermodynamic model for the cylinder and exhaust manifold of a natural gas 2-megawatt four-stroke internal combustion engine. *Heliyon* **2019**, *5*, e02700. [CrossRef] [PubMed]

37. Fan, W.; Han, Z.; Li, P.; Jia, Y. Analysis of the thermodynamic performance of the organic Rankine cycle (ORC) based on the characteristic parameters of the working fluid and criterion for working fluid selection. *Energy Convers. Manag.* **2020**, *211*, 112746. [CrossRef]

38. Iso, T.; Standards, I. *Environmental Management The ISO 14000 Family of International Standards ISO in Brief ISO and the Environment*; International Organization for Standardization: Geneva, Switzerland, 2009.

39. Valencia Ochoa, G.; Cárdenas Gutierrez, J.; Duarte Forero, J. Exergy, Economic, and Life-Cycle Assessment of ORC System for Waste Heat Recovery in a Natural Gas Internal Combustion Engine. *Resource* **2020**, *9*, 2. [CrossRef]

40. Michos, C.N.; Lion, S.; Vlaskos, I.; Taccani, R. Analysis of the backpressure effect of an Organic Rankine Cycle (ORC) evaporator on the exhaust line of a turbocharged heavy duty diesel power generator for marine applications. *Energy Convers. Manag.* **2017**, *132*, 347–360. [CrossRef]

41. ASHRAE. *ANSI/ASHRAE Standard 62.1-2010. Ventilation for Acceptable Indoor Air Quality*, 62nd ed.; American Society of Heating, Refrigerating, and Air-Conditioning Engineers, Inc.: Atlanta, GA, USA, 2010.

42. Ke, H.; Xiao, Q.; Cao, Y.; Ma, T.; Lin, Y.; Zeng, M.; Wang, Q. Simulation of the printed circuit heat exchanger for S-CO2 by segmented methods. *Energy Procedia* **2017**, *142*, 4098–4103. [CrossRef]

43. Ding, Y.; Liu, C.; Zhang, C.; Xu, X.; Li, Q.; Mao, L. Exergoenvironmental model of Organic Rankine Cycle system including the manufacture and leakage of working fluid. *Energy* **2018**, *145*, 52–64. [CrossRef]

44. Valencia Ochoa, G.; Piero Rojas, J.; Duarte Forero, J. Advance Exergo-Economic Analysis of a Waste Heat Recovery System Using ORC for a Bottoming Natural Gas Engine. *Energies* **2020**, *13*, 267. [CrossRef]

45. Valencia, G.; Duarte, J.; Isaza-Roldan, C. Thermoeconomic Analysis of Different Exhaust Waste-Heat Recovery Systems for Natural Gas Engine Based on ORC. *Appl. Sci.* **2019**, *9*, 4017. [CrossRef]

46. Valencia Ochoa, G.; Acevedo Peñaloza, C.; Duarte Forero, J. Thermoeconomic optimization with PSO Algorithm of waste heat recovery systems based on Organic Rankine Cycle system for a natural gas engine. *Energies* **2019**, *12*, 4165. [CrossRef]

47. Preißinger, M.; Brüggemann, D. Thermoeconomic Evaluation of Modular Organic Rankine Cycles for Waste Heat Recovery over a Broad Range of Heat Source Temperatures and Capacities. *Energies* **2017**, *10*, 269. [CrossRef]
48. Tchanche, B.; Lambrinos, G.; Frangoudakis, A.; Papadakis, G. Low-grade heat conversion into power using organic Rankine cycles-A review of various applications. *Renew. Sustain. Energy Rev.* **2011**, *15*, 3963–3979. [CrossRef]

MDPI

St. Alban-Anlage 66

4052 Basel

Switzerland

Tel. +41 61 683 77 34

Fax +41 61 302 89 18

www.mdpi.com

Energies Editorial Office

E-mail: energies@mdpi.com

www.mdpi.com/journal/energies